MAKING SENSE OF WORD PROBLEMS

CONTEXTS OF LEARNING
Classrooms, Schools and Society

Managing Editors:

Bert Creemers, *GION, Groningen, The Netherlands*
David Reynolds, *School of Education, University of Newcastle upon Tyne, England*
Sam Stringfield, *Center for the Social Organization of Schools, Johns Hopkins University*

MAKING SENSE OF WORD PROBLEMS

LIEVEN VERSCHAFFEL
University of Leuven, Belgium

BRIAN GREER
Queen's University Belfast, Northern Ireland

ERIK DE CORTE
University of Leuven, Belgium

SWETS & ZEITLINGER
PUBLISHERS

LISSE   ABINGDON   EXTON (PA)   TOKYO

Library of Congress Cataloging-in-Publication Data

Applied for

Cover design: Ivar Hamelink
Typesetting: Red Barn Publishing, Skeagh, Skibbereen, Co. Cork, Ireland
Printed in The Netherlands by Krips b.v., Meppel

ISBN 90 265 1628 2 Hardback

Contents

Preface

In this monograph we report and reflect upon a series of studies that have been done over the past few years to investigate the phenomenon of "suspension of sense-making" when doing word problems within the culture of mainstream mathematics education.

This book is the product of several years of intensive work by the three authors. Although none of the individual empirical studies reported in the book was carried out by all of us together, the overall research program, the design of the experiments and the theoretical analyses of the outcomes are the result of a truly collaborative effort. In the course of this collaboration, and in particular in working together on the writing of this book, we have met together on several occasions and we thank the British Council for their travelling grants to support these meetings.

We also thank the Fund for Scientific Research – Flanders for a three-year research grant (G.0149.96) which allowed the realization of several of the Leuven studies reported in the first and the second part of this book.

While the monograph focuses on the work that has been executed in our own two research centers, it also relies heavily on the work of scholars who initially aroused our interest in this phenomenon and others with whom we have had a close-knit and ongoing collaboration on this topic over the past few years, as evidenced in mutual visits, joint symposia presented at international conferences, and a special issue of the journal *Learning and Instruction*. In this respect, we particularly want to acknowledge the influence and work of Kurt Reusser, Fritz Staub, Rita Stebler, Roger Säljö, Jan Wyndhamn, Ernest Van Lieshout, Hajime Yoshida, and Claude Gaulin. We are also grateful to Koeno Gravemeijer, Jim Greeno, Gijoo Hatano, and Jeremy Kilpatrick, whose thoughtful comments as discussants in symposia and/or in the special issue of *Learning and Instruction* helped us greatly to refine our theoretical analysis.

We also want to thank several other scholars who have supported us in different ways in the realization of this book, namely: Fred Goffree, Swapna Mukhopadhyay, Jim Pellegrino, Alexander Renkl, Christoph Selter, Adri Treffers, and Marja Van den Heuvel-Panhuizen.

Many collaborators and students from our respective research groups have helped us carry out the empirical studies that are reported in the book. For the Leuven studies, our thanks are due to Hedwig Bogaerts, Inge Borghart, Dirk De Bock, Sabine Lasure, Elie Ratinckx, Griet Van Vaerenbergh, and Heidi Vierstraete. For the studies done in Belfast, our thanks are due to Lisa Caldwell, Marion Yamin-Ali, and Vivienne Boyle.

We also want to thank the series editors for inviting us to write this book and for their valuable comments on the concept and on the first draft of the book, the anonymous reviewers, and the publisher who worked hard to move this book quickly through to publication.

Finally, special thanks are due to Betty Van den Bavière and Karine Dens for the competence, patience, and care with which they have helped us in preparing the text for publication.

Lieven Verschaffel
Brian Greer
Erik De Corte

Introduction

Mathematics provides a set of tools for describing, analyzing and predicting the behavior of systems in different domains of the real world. This practical utility of mathematics has always provided one major justification for the important role of mathematics in the (elementary) school curriculum. In particular, the inclusion of application problems – traditionally in the format of word problems (or "verbal problems" or "story problems") – was, and is, intended to develop in students the skills of knowing when and how to apply their mathematics effectively in various kinds of problem situations encountered in everyday life and at work (e.g., Burkhardt, 1994).

It is not simple to provide a precise and complete definition of "word problem" (Semadeni, 1995). Word problems can be defined as verbal descriptions of problem situations wherein one or more questions are raised the answer to which can be obtained by the application of mathematical operations to numerical data available in the problem statement. In their most typical form, word problems take the form of brief texts describing the essentials of some situation wherein some quantities are explicitly given and others are not, and wherein the solver – typically a student who is confronted with the problem in the context of a mathematics lesson or a mathematics test – is required to give a numerical answer to a specific question by making explicit and exclusive use of the quantities given in the text and mathematical relationships between those quantities inferred from the text.

According to this definition, a characteristic feature of word problems is the use of words to describe a (usually hypothetical) situation. The problems "Pete wins 3 marbles in a game and now has 8 marbles. How many marbles did he have before the game?" and "One kilogram of coffee costs 12 EUR. Susan buys .75 kilogram of coffee. How much does she have to pay?" are typical examples of simple word problems, while $8 - 3 = ?$ and $12 \times .75 = ?$ are not.

The definition might seem to include tasks such as "What do you get if you subtract 3 from 8?" and "What number do you obtain when you multiply 12 by .75?", but we agree with Semadeni's (1995) position that a word problem should refer to an existent or imaginable meaningful context, excluding the context of doing a purely numerical calculation. Thus, verbally stated numerical problems (like the last two cited above) are not considered as word problems.

Word problems are not necessarily problems in arithmetic, but could be concerned with other branches of mathematics, particularly algebra (Reed, 1999), but also geometry, logic, probability, and so on. Our main focus throughout is on arithmetical word problems, although these merge into algebraic problems. For instance the first problem cited above ("Pete wins 3 marbles in a game and now has 8 marbles. How many marbles did he have before the game?") may be construed as an arithmetic problem the solution of which is found by subtracting 3 from 8, or it might be construed as an algebra problem requiring the solution of the equation $x + 3 = 8$. As Reed (1999, p. 76) points out, it is not the task per se that determines whether or not it is an algebra problem, but rather the approach used to solve the problem.

Another question is whether a word problem must, by definition, be given in exclusively written form. Like Semadeni (1995), we take the view that this is not the case. Thus, word problems can take the form of a combination of written text and other kinds of information (e.g., a table, a picture, a drawing, a video...) and can also be presented orally (implying the use of intonation, gesture, and other non-verbal forms of communication). Furthermore, while people are, of course, frequently confronted with quantitative tasks and dilemmas in real-life settings out of school, we do not use the term "word problem" to refer to these out-of-school situations. Word problems are associated with a school setting, and, more specifically, with a setting wherein a learner is being asked to solve such a problem in the context of a mathematics lesson or a mathematics test.

Finally, according to our definition the presence of the word "problem" in the term "word problem" does not involve any assumption about the level of difficulty or problematicity of the task. In other words, a word problem does not necessarily constitute a problem (in the cognitive-psychological sense of the word) for a particular student, and consequently does not necessarily require the use of higher-order thinking and problem-solving skills. Whether a word problem deserves the label of a "problem" depends on the relationship between the concepts and skills that are required to produce a satisfactory answer, on the one hand, and the solver's knowledge, skills, beliefs, attitudes, etc., on the other hand. As will be documented throughout this book, one of the major shortcomings of current mathematics education, in our view, is that word problems are almost invariably presented to students as tasks that should be recognized as familiar and solved by relying on previously learned formulas and procedures, rather than – at least sometimes – problems that they may not immediately recognize as belonging to a particular problem type, for which they have no ready-made solution, and which require higher-order thinking.

Several components can be distinguished in a word problem (Goldin & McClintock, 1984; Silver, 1985; Verschaffel, 1984):

- The mathematical structure: i.e., the nature of the given and unknown quantities involved in the problem, as well as the kind of mathematical operation(s) by which the unknown quantities can be derived from the givens.
- The semantic structure: i.e., the way in which an interpretation of the text points to particular mathematical relationships – for example, when the text implies a change from an initial quantity to a subsequent quantity by addition or subtraction, or a combination of disjoint subsets into a super-set, or the additive comparison between two collections, then in each case operations of addition or subtraction are indicated.
- The context: What the problem is about, e.g., whether, in the case of an additive problem involving combination of disjoint sets it deals with groups of people joining each other, with combining collections of objects, etc.

- The format: i.e., how the problem is formulated and presented, involving such factors as the placement of the question, the complexity of the lexical and grammatical structures, the presence of superfluous information, etc.

Over the past decades numerous studies have analyzed the role of these different kinds of task variables on the difficulty of problems, on the kind of strategies students use to solve these problems, and on the nature of their errors (for comprehensive reviews of these studies see, e.g., Fuson, 1992; Greer, 1992; Riley, Greeno, & Heller, 1983; Staub & Reusser, 1995; Verschaffel & De Corte, 1993, 1997a).

Why does school mathematics include word problems? Perhaps simply because they are there, and have been for many centuries. Indeed, word problems have always constituted an important part of the mathematics program in the (elementary) school. Historically their role in mathematics education dates back even to antiquity. One can find verbal problems already almost 4000 years ago in Egyptian papyri. They also figure in, for example, ancient Chinese and Indian manuscripts (e.g., Colebrooke, 1967; Libbrecht, 1973) as well as in arithmetic textbooks from the early days of printing, such as the Treviso arithmetic of 1478 (Swetz, 1987) and they continue to fill current mathematics textbooks. Despite this continuity across time and cultures, there has been little explicit discussion of why word problems should be such a prominent part of the curriculum and recent writers have called for a re-examination of the rationale for this privileged position (e.g., Gerofsky, 1996; Lave, 1992). We may infer that word problems have been included with the ostensible aim of accomplishing several goals, including the following:

- First, and most importantly, to offer practice for the situations of everyday life in which mathematics learners will need what they have learned in school (e.g., Thorndike, 1922). The tools of mathematics are powerful but, as with all tools, only if one knows how to use them across a range of suitable situations (Lave, 1992). The inclusion of word problems in school is designed to develop the skills to do this (= application function).
- To motivate students. Word problems may also be a means of convincing students that they really will need the mathematics they are learning in school when they grow up and move into the world. Through experience in doing these problems, students should become confident that once they have learned the techniques, they will be able to apply them in the future not only in school but also out of school in real situations (= motivation function).
- To evaluate the intelligence or mathematical ability of youngsters who will grow up to take different economic and social positions (= selection function).
- To train students to think creatively and/or to develop their heuristic skills and their problem-solving abilities (= thought-provoking function).
- To develop new mathematical concepts and skills. Carefully selected and sequenced word problems can provide a context for the exploration and

construction of new mathematical concepts, techniques, etc. (= concept-formation function).

Critics such as Lave (1992) and Gerofsky (1996) have cast doubt on the extent to which word problems, as presently used in schools, achieve these aims. Moreover, Gravemeijer (1997, p. 390) points to another implicit aim, namely computational practice, since "[m]ost text-book word problems are nothing more than poorly disguised exercises in one of the four basic operations". In the course of this book we will add to these criticisms and suggest how the aims listed above, and others, could be better achieved.

The application of mathematics to solve problem situations in the real world, otherwise termed mathematical modeling, can be usefully thought of as a complex process involving a number of phases: understanding the situation described; constructing a mathematical model that describes the essence of those elements and relations embedded in the situation that are relevant; working through the mathematical model to identify what follows from it; interpreting the outcome of the computational work to arrive at a solution to the practical situation that gave rise to the mathematical model; evaluating that interpreted outcome in relation to the original situation; and communicating the interpreted results. As several authors have stressed, this process of solving mathematical application problems has to be considered as cyclic, rather than as a linear progression from givens to goals (Burkhardt, 1981, 1994; Greer, 1997, Lesh & Lamon, 1992a).

Figure 0.1 schematically represents this way of characterizing the process of solving mathematical application problems. This characterization can also be used to describe the process of solving word problems construed as being a special kind of application problem presented in a particular (i.e. verbal, or at least predominantly verbal) format and in a particular (i.e. instructional) setting (Burkhardt, 1981, 1994; Nesher, 1980). While the two rightmost boxes in Figure 0.1 refer to the "purely" mathematical aspects of the process, the two leftmost boxes deal with its involvement with the real world.

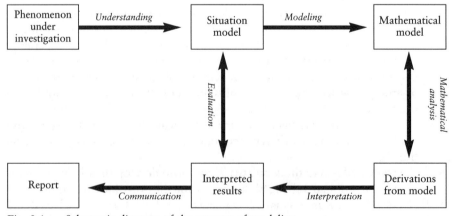

Fig. 0.1   Schematic diagram of the process of modeling.

As indicated earlier, the central theme of this book is that word problems should be problematised. By this statement we mean, firstly, that their characteristics, use and rationale should be analysed and re-evaluated (Gerofsky, 1996; Lave, 1992). Secondly, in challenging the predominant way in which they are used currently in mathematics classrooms as routine applications without judgment or any higher-level thinking skills, we present an alternative conceptualisation (consistent with the suggestions of many others) whereby they are treated as application problems, simple exercises in mathematical modeling.

Despite its prominence as a theme in mathematics education, the term "mathematical problem solving" is hard to define precisely. It is useful to distinguish two aspects, the cognitive-psychological processes and the nature of the mathematical tasks presented. A very general definition, from a cognitive-psychological perspective, is that a problem exists for a person when (s)he is faced by a presented task for which no standard procedure is known (see, e.g., Kilpatrick, 1985). To succeed on such a task, therefore, some creativity is implied in terms of restructuring available knowledge.

Mathematical tasks tend to be either labeled or not labeled as problems depending on how amenable they are to standard techniques (e.g., solving simultaneous equations would not normally be labeled as a problem, though for students at a certain stage it can be). However, whether a particular task confronted by a particular individual constitutes a problem in the psychological sense, depends on the relationship between the task and the individual's current cognitive repertoire, as indicated earlier.

From a mathematical point of view, the tasks set as potential problems may be classified in terms of the branch of mathematics (arithmetic, algebra, geometry, logic, probability, etc). A second major distinction (not hard and fast) is between pure and applied tasks. For example, the problem of proving Fermat's Last Theorem (that $x^n + y^n = z^n$ has no solutions with x, y, z, n all positive integers for n > 2) is "pure" in the sense that it deals exclusively with the abstract structure of the number system. By contrast, predicting the path taken by a projectile is "applied" in the sense that it deals, albeit in idealised form, with the physics of the real world. (It is also a good example of mathematical modeling since the results of the model depend on the assumptions built in – for example, whether or not air resistance is taken account of).

In terms of mathematical problem solving in general, there is now a broad consensus that successful solution depends on the simultaneous and integrated application of several components such as a well-organized and flexibly accessible knowledge base, heuristic methods, metacognitive processes, and affective aspects. Moreover, expertise in mathematical problem solving requires not just the mere sum of these four components. Rather, they should be applied integratively and complemented by some additional aptitudes that are not clearly covered by the four components as described above. Examples are students' *inclination* to use their knowledge and skills whenever appropriate and their *alertness* and *sensitivity* for contexts and situations in which it is appropriate to do so (De Corte, 1995; De Corte, Greer, & Verschaffel, 1996). To capture this aspect, the term "mathematical disposition" has become more and more in vogue as, for instance, in the following quotation from the *Curriculum and Evaluation Standards for School Mathematics* produced by the National Council of Teachers of Mathematics (NCTM) in the United States:

> Learning mathematics extends beyond learning concepts, procedures, and their applications. It also includes developing a disposition toward mathematics and seeing mathematics as a powerful way for looking at situations. Disposition refers not simply to attitudes but to a tendency to think and to act in positive ways. Students' mathematical dispositions are manifested in the way they approach tasks – whether with confidence, willingness to explore alternatives, perseverance, and interest – and in their tendency to reflect on their own thinking. (National Council of Teachers of Mathematics, 1989, p. 233)

The same analysis, with appropriate changes in emphasis, may be applied to application problems, and, in particular, to word problems conceptualised as such.

- A well-organized and flexibly accessible knowledge base is required. Clearly, one component of this knowledge base is the combination of facts, symbols, algorithms, concepts, rules etc. that constitute the basic elements of mathematics. Besides this purely mathematical knowledge, successful performance on word problems obviously requires linguistic knowledge. Further, for particular word problems, real-world knowledge about the concrete contexts or situations described (buses, shopping, social interactions between people etc.) is indispensable for selecting those aspects of the problem situation that need to be included and what are appropriate mathematical ways of representing relationships between those aspects in a model, and for interpreting and evaluating the implications derived from this model in terms of the original problem situation (Burkhardt, 1981, 1994; Greer, 1997; Lesh & Lamon, 1992a).

- The importance of heuristic methods is not as obvious as in the types of pure mathematical problems analysed and investigated by, for example, Polya (1957) and later Schoenfeld (1985). Nevertheless, many heuristics are applicable, such as visualizing the situation described by producing a diagram or drawing, decomposing a problem into subproblems, considering an analogous but simpler problem, beginning with a very simple model of the situation.

- Metacognition is variously defined (e.g., Schoenfeld, 1987a), but for our present purposes, we will take it as having two major components. The first comprises knowledge and beliefs concerning one's own cognitive functioning (e.g., believing that one's mathematical ability is strong, or that one cannot remember mathematical formulas). The second comprises skills and strategies for the regulation of one's cognitive processes (e.g., monitoring an ongoing solution process; evaluating, and, if necessary, revising a model; reflecting on one's modeling activities).

- Affective components, involving beliefs about mathematics (e.g., believing that a word problem can be solved by applying one of the four basic operations to the two numbers mentioned in the problem and that some cue in the problem will indicate which is the appropriate operation), attitudes and emotions (e.g., liking or disliking word problems).

For several years it has been argued by many mathematics educators that the current practice of word problems in school mathematics does not at all foster in students a genuine disposition towards mathematical modeling. A major criticism in this respect, which will be the focus of the present monograph, is the following. Rather than functioning as realistic and authentic contexts inviting or even forcing pupils to use their common-sense knowledge and experience about the real world in the different stages of the process of solving mathematical application problems, school arithmetic word problems are perceived as artificial, puzzle-like tasks that are unrelated to the real world (Cooper, 1994; Davis, 1989; De Corte & Verschaffel, 1985; Freudenthal, 1991; Greer, 1993; Kilpatrick, 1987; Nesher, 1980; Nunes, Schliemann, & Carraher, 1993; Reusser, 1988; Säljö, 1991; Schoenfeld, 1991; Silver, Shapiro, & Deutsch, 1993; Treffers, 1987; Verschaffel & De Corte, 1997a).

In this monograph we report and reflect upon a series of studies that have been done over the past few years to investigate this phenomenon of "suspension of sense-making" (Schoenfeld, 1991) when doing word problems within the culture of mainstream mathematics education. While the focus will be on studies that have been executed in our own two research centres, we also rely on the work of the other scholars who aroused our interest in this phenomenon originally and those with whom we have had a close-knit and ongoing collaboration on this topic over the past few years.

Generally speaking, the work that will be reported in these chapters can be considered as an additional line of evidence for the importance of the social and cultural context in which mathematical thinking and learning take place (De Corte et al., 1996). Another well developed line of research comprises the studies of "everyday cognition" (e.g., of street vendors, supermarket shoppers, dairy-workers) by scholars such as Scribner (1984), Lave (1988), Carraher, Carraher, and Schliemann (1985) and others. A first major outcome of these studies on everyday cognition is that people are remarkable efficient in dealing with quantitative problems encountered in their everyday professional and social activities as compared with the school mathematics context. Second, these studies show how the goals and conditions of the practical activity lead to the use of non-formal mathematical reasoning and computational processes that differ considerably from the formal, standardized procedures typically transmitted in school. Taken as a whole, the work of these ethnomathematicians has led to a clearer recognition of the differences between students' activities and experiences within school mathematics and their out-of-school activities and experiences, and especially of the lack of transfer of knowledge and skills acquired in the first context to the second one. Indeed, subjects in the above-mentioned studies notably did not spontaneously and efficiently apply their formal mathematical knowledge and skills learned at school in out-of-school contexts. These findings contradict, accordingly, the optimistic expectation that people should apply knowledge and skills acquired in a formal school setting to relevant situations in everyday life.

In this monograph we examine the other side of this particular coin. That is to say, rather than looking at school mathematics not being taken into real life and work situations, we focus on how real-world knowledge is not brought into – or, rather, *not allowed* to be brought into – school mathematics. Our in-depth analysis of this remarkable phenomenon will further reveal that word problem activity at school takes place within a very special type of context that influences how students learn and think, and that complicates and hinders transfer from what is learnt in school to out-of-school contexts. This is completely in line with Lave's (1988) insight that, rather than viewing school mathematics as abstract and formal, and therefore unsituated and universally transferrable, it constitutes, rather, another particular form of situated activity. Or, as Lave (1992, p. 81) put it "math in school is situated practice: school is a site of specialized everyday activity – not a privileged site where universal knowledge is transmitted".

From this perspective, this monograph can be considered as an attempt to further document and understand the contextual dependency of human cognition – a major theme in cognitive and instructional psychology today (Greeno & Middle School Mathematics Through Applications Project Group, 1998; Lave, 1988, 1992; Rogoff, 1984; Rogoff & Lave, 1984; Säljö & Wyndhamn, 1987).

The monograph is divided into three parts. Part 1 reviews the basic observations and research findings with respect to the phenomenon of "suspension of sense-making" without going in any depth into interpretations of those observations or into our emerging theoretical ideas. In the first chapter of Part 1 we describe some earlier examples of striking evidence of students' lack of sense-making in school mathematics, which aroused our interest in the topic. This work was the basis for the studies by Greer (1993) and by Verschaffel, De Corte, and Lasure (1994) with students in Northern Ireland and Belgium, respectively, which provided additional and more systematic evidence that after several years of traditional mathematics instruction students have developed a tendency to reduce word problem solving to selecting what they take to be the correct arithmetic operation with the numbers given in the problem, without seriously taking into account their common-sense knowledge and realistic considerations about the problem context. These two studies, as well as some replications with students in other countries, are discussed in Chapter 2. In the third and fourth chapters we then present two related but distinguishable lines taking the research further. While the first line of research (covered in Chapter 3) investigated the effects of several forms of scaffolding aimed at enhancing students' alertness to the problematic nature of the problems used in these two original studies and their replications, the second (reviewed in Chapter 4) looks at the effectiveness of attempts to increase the authenticity of task presentation.

Having presented and discussed in Part 1 the available studies documenting and probing the phenomenon of students' tendency to tackle arithmetic word problems with apparent suspension of sense-making, Part 2 reflects the broadening of both our theoretical perspective and our experimental program. Chapter 5 begins with a detailed analysis of the way in which word problems are currently taught in typical mathematical classrooms, drawing on a substantial body of observations, experimental evidence and theoretical reflection. Having discussed the strongly shaping influence of textbooks and assessment instruments, we look at the role of the most important factor, namely the teacher. More specifically, a study of pre-service teachers' performance on the same word problems given to students in the studies discussed earlier, and of their views about these problems, is reported. This chapter forms the transition to the second chapter of Part 2, Chapter 6, in which a number of studies are reported that have gone beyond ascertaining studies (i.e., studies restricted to documenting the existing state of affairs). These design experiments illustrate how by immersing students in a fundamentally changed learning environment they can acquire what we consider to be more appropriate conceptions about, and strategies for doing, word problems. A common feature of the approach followed in these design experiments is that word problems are not conceived as artificial, puzzle-like tasks that can always be unambigously solved by performing one (or a combination) of the four basic arithmetic operations with the given numbers, but as genuine exercises in realistic mathematical modeling. Bearing in mind the influential role played by assessment in shaping instructional practice, and in signaling to both students and teachers what is valued in mathematics, concomitant attempts to improve the quality of assessment are also reviewed, and a study showing effects on teaching of reforming assessment is summarised.

Having discussed the empirical evidence in the first two parts, the third part turns to a wider discussion of theoretical issues, a further analysis of the features of the (mathematics) educational system that lie at the roots of what we consider to be the outcomes seriously detrimental to many students' understanding and conception of mathematics, and suggestions for rethinking the role of word problems within the curriculum. Having reflected upon the nature of the mathematical modeling process in Chapter 7, we return, in Chapter 8, to the nature of word problems, reviewing how they have been used in mathematics education across time and cultures. They are considered from multiple perspectives – historical, cultural, linguistic, and sociological. In Chapter 9 we end the book with suggestions for reconceptualising the role that word problems might play in mathematics education and how the approach to teaching them might be changed accordingly.

Part 1

Suspension of Sense-making: Documenting the Phenomenon

1

Examples of Suspension of Sense-making in the Mathematics Classroom

When Flaubert was a very young man, he wrote a letter to his sister, Carolyn, in which he said: "Since you are now studying geometry and trigonometry, I will give you a problem: A ship sails the ocean. It left Boston with a cargo of wool. It grosses 200 tons. It is bound for Le Havre. The mainmast is broken, the cabin boy is on deck, there are 12 passengers aboard, the wind is blowing East-North-East, the clock points to a quarter past three in the afternoon. It is the month of May. How old is the captain?" (Wells, 1997, p. 234, also cited in the original French by Baruk, 1985, p. 114)

In this first chapter we briefly describe a number of striking examples of students' lack of sense-making in school mathematics that we have encountered in the literature and that formed the basis of the studies by Greer (1993) and Verschaffel et al. (1994) that are presented in Chapter 2. While some of these examples are rather anecdotal, others are supported by intensive and systematic empirical research.

How old is the captain?

In the late seventies and early eighties, some spectacular illustrations of the failure of pupils to make proper use of real-world knowledge and sense-making

faculties in their solutions of school word problems were provided by French
and German researchers.

A researcher from Grenoble (Institut de Recherche sur l'Enseignement des
Mathématiques de Grenoble, 1980) presented a group of children from the first
and the second grade of elementary school the following absurd problem: "There
are 26 sheep and 10 goats on a ship. How old is the captain?" and found that the
vast majority of them were prepared to offer an answer to this unsolvable problem by combining the numbers given in the problem to produce answers without
apparent awareness of the meaninglessness of the problem and of their solution.

Shortly afterwards, the same Institut de Recherche sur l'Enseignement des
Mathématiques (IREM) replicated this experiment on a larger scale. Six variations of the problem about the captain's age, for example "There are 125 sheep
and 5 dogs in a flock. How old is the shepherd?" and "I have 4 lollies in my right
pocket and 9 sweeties in my left pocket. How old is my father?" were individually administered in a written form to large groups of children from the first
and second grade of elementary school (7–9-year-olds) and from the middle
grades (9–11-year-olds). Each problem was followed by the question "What do
you think of the problem?". Only 20 of the 171 children aged 7 to 9 (12%), and
74 of the 118 children aged 9 to 11 (62%) reacted by saying they could not
respond properly to the questions. All the others performed some arithmetical
operation on the given numbers without expressing any doubt. One pupil's pro-
tocol for the shepherd problem mentioned above went like this: "$125 + 5 = 130$
... this is too big, and $125 - 5 = 120$ is still too big ... while ... $125 \div 5 = 25$
... that works ... I think the shepherd is 25 years old".

In the first chapter of her book *L'âge du capitaine*, which is mainly devoted
to this intriguing phenomenon, Baruk (1985) expressed her feelings of concern,
and even of indignation, about the fact that so many "normal" children showed
such irrational and abnormal behavior by adding sheep and goats in order to
work out the captain's age in the first try-out of this problem:

> Voilà. Vous avez bien lu. Des enfants qui sont comme vous et moi,
> c'est-à-dire comme ceux que nous avons été ou comme ceux qui sont
> les nôtres, des petits Français du dernier quart du 20e siècle qui ne
> sont ni en IMP ni en IMPP, ni en hôpital de jour, ni en hôpital psy-
> chiatrique, des enfants donc "normaux", et destinés à devenir les
> citoyens de l'an 2000, pour obtenir l'âge du capitaine ont accouplé
> des moutons et des chèvres. (Baruk, 1985, p. 23)

Later in the chapter, Baruk (1985, p. 26) commented as follows on the unac-
ceptably high percentages of 7–9 and 9–11-year-old pupils responding to the set
of six similar absurd problems in a stereotyped and uncritical way:

> ... au lieu du formidable éclat de rire général, du "mais ils sont fous
> ces profs!" qui auraient unanimement dû acceuillir ces énoncés
> immédiatement recevables comme insensés, ou comme irrecev-
> ables, on obtient ça: un refus poli d'un peu plus de 1 élève sur 10

au Cours élémentaire, d'un peu plus de la moitié au Cours moyen. Et pour les autres, tous les autres, cette terrifiante acceptation de l' inacceptable ...

Radatz (1983, 1984) conducted a test on more than 300 German children from Kindergarten to Grade 5 in which, amongst normal school word problems, some unsolvable problems were given such as "Katja invites 8 children to come to her birthday party, which takes place in 4 days. How old will Katja be on her birthday?". Contrary to the findings from the IREM, Radatz observed that the percentage of children trying to reach some solution increased with years of schooling. While only about 10% of the Kindergarten children and the first-graders worked on these unsolvable problems, the percentages of second-graders (30%) and third- and fourth-graders were much higher (about 60%), before going down only slightly to 45% in Grade 5. From this remarkable developmental trend, Radatz concluded that pupils' solution behavior is strongly influenced by the amount of mathematics teaching they have received. Children with little mathematical experience try to analyze the problem more carefully, whereas older students have a specific attitude towards mathematics – they have learned to see mathematics as a kind of play with artificial rules and without any link to out-of-school reality (Radatz, 1983). (For some methodological criticisms of Radatz' study and a replication study with conflicting results, see Stern (1992).)

Reusser (1988) observed Swiss pupils from Grade 1 through 5 working on some of the IREM problems in a typical school mathematics setting. As in the French and German studies, the vast majority of the pupils in Reusser's study attempted to provide numerical answers to the problems. Their overwhelming tendency was, once again, to ask themselves whether to add, subtract, multiply, or divide rather than ask whether the problem made sense. In an article entitled *Mathematics as sense-making: An informal attack on the unfortunate divorce of formal and informal mathematics*, Schoenfeld (1991, p. 316) comments as follows on Reusser's observations:

> The students he interviewed not only failed to note the meaninglessness of the problems as stated, but went ahead blithely to combine the numbers in the problem and produce answers. They could do so by engaging in what might be called suspension of sense-making – suspending the requirement that the way in which the problems are stated makes sense.
>
> . . . There is reason to believe that such suspension of sense-making develops in school, as a result of schooling.

Several authors questioned the generality of these spectacular findings and conducted their own versions of these experiments. These skeptics included Bransford and Stein (1993, p. 196), who wrote about their experience as follows:

Our reaction to Reusser's data was that this must have been a special group of students who had been poorly taught. We gave the
sheep-and-goat problem to one of our own children who was in fifth
grade. Much to our surprise, and dismay, the answer given was 36.
When we asked why, we were told: "Well, you need to add or subtract or multiply in problems like this, and this one seemed to work
the best if I add".

In sum, elementary school pupils with some years of experience with school
arithmetic often seem to react to absurd problems like the one about the captain's age by somehow operating on the numbers in the problem to arrive at a
numerical answer. The authors cited above consider these bizarre answers as ter-
rifyingly strong evidence that, after a couple of years of experience with traditional mathematics education, students approach word problems in a
thoughtless and mechanical way, without paying attention to the context and
without any reference to their common sense. Throughout this book, we will
come back regularly to this extreme case of "suspension of sense-making", providing some further observations and interpretations which offer a somewhat
different perspective on this phenomenon (see, e.g., Brissiaud, 1988; Brousseau,
1984, 1990, 1997; Selter, 1994; Stern, 1992).

How many buses are needed?

Another oft-cited and extensively analyzed example is the following, used for
the first time with a stratified sample of 13-year-olds in the Third National
Assessment of Educational Progress in the U.S. (Carpenter, Lindquist,
Matthews, & Silver, 1983): "An army bus holds 36 soldiers. If 1128 soldiers
are being bussed to their training site, how many buses are needed?". Of 70%
of the students who correctly carried out the division 1128 by 36 to get a quo-
tient of 31 and remainder of 12, only 23% (of the total number of students)
gave the answer as 32 buses, 19% gave the answer as 31 buses, and 29% gave
the answer as "31, remainder 12". A similar division problem appeared on the
1983 version of the California Assessment Program (CAP) Mathematics Test for
Grade 6, and was answered correctly by about 35% of the sixth-graders in
California (Silver, 1986; cited in Silver et al., 1993). As in the NAEP assessment,
most students erred by giving non-whole-number answers or rounding the outcome of the division to its nearest whole-number predecessor.

To better understand the solution processes and interpretations underlying
the observed difficulty that students have in solving division problems with
remainders (DWR), Silver and associates conducted several investigations in
which, unlike the multiple-choice format used in the earlier research, a free-
response paper-and-pencil format was used.

In one study (Silver et al., 1993) the sample consisted of 195 students from
grades 6–8. The following task was administered to each student in the sample:

"The Clearview Little League is going to a Pirates game. There are 540 people, including players, coaches and parents. They will travel by bus, and each bus holds 40 people. How many buses will they need to get to the game?" Besides 540, two other numbers of people were used so that student responses were obtained for DWR problems with remainder sizes equal to one-half (540), less than one-half (532), and greater than one-half (554). The reason for using these three versions of the problem was the authors' hypothesis that the number of "leftovers" might influence students' situation-based interpretations and final solutions. The instructions accompanying the task directed students to show their work, to place their answer in the answer space provided, and to explain their answer in writing.

A detailed and integrated inspection of students' (a) solution processes, (b) numerical answers, and (c) interpretations revealed that approximately 78% of all solution processes could be unambiguously categorized into one of the following two (hypothetical) solution models:

- A successful, semantically-based solution (SS) model, wherein the problem solver builds a semantic representation of the story situation starting from the problem text, then maps it into a mathematical model representation, then performs the required computation(s) within the referential system of mathematics, and finally maps the computational result back to the original story (text) representation (see Figure 0.1).
- An unsuccessful (US) model, wherein the student maps successfully from the problem text into a mathematical model, and computes an answer within the domain of the mathematics model, but fails afterwards to return to the problem story text or the story situation referent in order to determine the best answer to the question. This corresponds to omitting the interpretation and evaluation steps in the model presented in Figure 0.1.

While direct supporting evidence for the hypothesized SS model was provided by approximately 32% of the responses, the US model was supported by approximately 46% of the responses. (The remaining 22% responses could not be unambiguously placed into either of these two categories).

Students whose responses were consistent with the first model (SS) either gave the numerical answer of 14 and provided an appropriate interpretation (26%), provided an answer other than 14 but also gave an appropriate interpretation (3%), or had flaws in the execution of their solution procedure but were able to obtain an answer of 14 and give an appropriate interpretation (3%). Students whose responses were consistent with the second model (US) correctly executed an appropriate solution procedure but provided no interpretation for their incorrect numerical answer (22%), or incorrectly executed an appropriate procedure, gave a numerical answer other than 14, and provided no interpretation (24%). The written responses of the students offered no direct evidence of students being influenced by the size of the remainder in interpreting their solutions or arriving at their final numerical answer.

Overall, the study of Silver et al. (1993) provided considerable evidence that middle school students experienced great difficulties in solving DWR problems, and that computational requirements were not the major barrier to obtaining a correct solution but rather that unsuccessful solutions were more often due to students' omitting the interpretation and evaluation steps to make sense of their computational results. The finding that students' performance was adversely affected by the dissociation of sense-making from the solution of what is a typical school mathematics problem points, according to these authors, to the need for more instructional attention to sense-making as a part of school mathematics instruction (Silver et al., 1993).

In the context of one of the major issues of cross-national comparisons of mathematics achievement, namely whether Asian students' greater proficiency on mathematical tasks is restricted to fairly routine computational tasks or generalizes to tasks requiring higher forms of mathematical thinking and problem solving, Cai and Silver (1995) compared the results of the students from the study of Silver et al. (1993) with those of a comparable group of 186 Chinese students (fifth and sixth graders), judged by their teachers to be of average ability in mathematics. Each subject received a booklet containing a set of seven tasks, one of which was the following DWR problem: "Students and teachers at Guangming Elementary School will go by bus for spring sightseeing. There is a total of 1128 students and teachers. Each bus holds 36 people. How many buses are needed?" Each student's written response to the DWR problem was subjected to a fine-grained analysis, which was done in the same way as in Silver et al.'s (1993) study. The major outcome of this study was that although the Chinese students outperformed the U.S. students in the (correct) execution of the division computation procedure (more than 80% of the Chinese students correctly executed the division algorithm, versus only 61% of the U.S. students), that superiority did not carry over to better performance on the DWR problem. In fact, the percentage of Chinese students (30%) who gave the most appropriate numerical answer to the DWR problem (namely 32 buses) was lower than was reported in the study by Silver et al. (1993), in which 43% of the U.S. students gave the corresponding most appropriate answer (namely 14 buses) for the similar, but computationally simpler, DWR problem. Also, the percentage of Chinese students providing an appropriate written explanation or interpretation (10%) was considerably lower than for the U.S. students (about 33%). These findings indicate that the lack of sense-making when solving DWR problems is not restricted to U.S. students, but applies also to students from another country, one typically found to have superior achievement for mathematics compared with the U.S.

How much to post a letter?

Säljö and Wyndhamn (1990) administered the everyday problem of establishing the postage rate for a letter that weighs 120 grams, using a letter-scale and a postage table, in the context of a mathematics lesson. This table was a copy of

Letters Domestic Regular letters (and picture postcards)	
Weight (in g) (not exceeding)	Postage
20	2.10
100	4.00
250	7.50
500	11.50
1000	15.50

Fig. 1.1   Excerpt from a Swedish postage table.

the official postage table of the Swedish Post. The critical part of the table is shown in Figure 1.1.

The participants, 45 pupils aged 12–13 years in the sixth form of a Swedish comprehensive school, co-operated in groups of three that were homogenized with respect to academic ability.

The results can be summarized as follows. First, the problem of establishing the postage rate for a letter that weighed 120 grams turned out to be quite a challenge for the participants, and only one-third of the groups eventually ended up with the correct postage rate (7.50 kroner). Second, all groups drew on school knowledge when solving the task, but this in no way was linked to finding the correct answer. For instance, the most common suggestion with which pupils came up during the solution process was 6.10 kroner, which was arrived by adding the postage rates for letters weighing 20 grams (2.10 kroner) and 100 grams (4 kroner) respectively. Alternative formal computational strategies represent ways to arrive at a postage rate by using a strictly proportional approach to calculate the postage per gram, and then to multiply by 120 or to multiply the upper limit of the first interval (20 grams) by 6. These computation-based solutions were suggested more often than those based on reading the table directly without performing any computation. Third, inappropriate solutions and procedures for handling the table appeared just as frequently in all ability groups as suggestions at some stage of the group work, but they were generally rejected during the longer and more elaborate interpretative work among the pupils of high academic ability. In fact, while four of the five groups of high achievers finally arrived at the correct postage rate, only one of the groups of average and none of the low performers did so. Taken as a whole, the study of Säljö and Wyndhamn (1990) can be considered as a good example of how children "situate their math practice differently in different settings, and, more specifically, of how quantitative terms and relations involved in word problems are assembled and transformed by methods quite different from those used in out-of-school settings" (Lave, 1992, p. 78).

Other examples of lack of sense-making in the mathematics classroom

In a caustic critical analysis of the culture of school mathematics, Davis (1989) reported the following classroom observation. In an introductory lesson about division, pairs of pupils were given five balloons that had to be shared between two people. One boy took scissors and cut in half the fifth balloon. Davis (1989, p. 144) asked in this respect: "Was this boy really thinking about solving the actual problem – i.e., effectively sharing the five balloons – or was he trying to accommodate himself to the peculiar tribal culture of the American school?"

In her theoretical analysis of the stereotyped nature of school word problems, Nesher (1980) gave the following striking illustration of pupils' inability to employ their own knowledge about the context in question when it is presented to them in the format of a school word problem and in the context of a mathematics lesson. She gave the following problem to fifth-graders "What will be the temperature of water in a container if you pour 1 jug of water at 80°F and 1 jug of water at 40°F into it?", and observed that the answer given by many children was "120°F"! If, according to Nesher (1980), you gave the same children the parallel problem "What will the water be like if you pour hot water and cold water into one container?" their answer will always be "You get lukewarm water". So, although the children had experience with the context presented to them in the school word problem, it was not evident in their response to that school problem. Rather, their problem-solving activity seemed to be dominated by a simple rule they have learned at school, namely that "when you put things together, you add their numbers" (Nesher, 1980).

In their survey of the research on students' reactions to verbal arithmetical problems with missing, surplus, and contradictory data, Puchalska and Semadeni (1987) refer to two studies with so-called "pseudoproportionality" problems. Markovitz, Hershkowitz, and Bruckheimer (1984) investigated how students solve pseudoproportionality problems where there are no exact logico-mathematical relations between the known and unknown elements and therefore no conclusion follows, e.g., a problem wherein the height of a 10-year-old boy was given and the question concerned what his height would be when he was 20. Many answers based on direct proportionality (i.e., doubling the boy's height because 20 is two times 10) were given. Another kind of pseudoproportionality problem – resembling Säljö and Wyndhamn's (1990) postage problem discussed above – was used by Bender (1985) in a study with upper elementary school children. The problem was: "A postage stamp for a standard letter from Aachen to Munich costs 60 pfennig. The distance from Aachen to Munich is 600 kilometers. The distance from Aachen to Frankfurt is 300 kilometers. How much is the postage for a standard letter from Aachen to Frankfurt?" In contrast with the preceding example about the boy's height, this question did have a unique and precise answer, provided one is aware that the cost of mailing an ordinary letter in Germany is the same for all pairs of cities. However, many pupils answered "30 pfennig".

At the beginning of an article about problem formulating, Kilpatrick (1987) asks the reader to consider an (imaginary) teaching/learning activity around the problem given in Figure 1.2, which was developed for use by teachers as the basis for a discussion on formulating problems. The problem is about the comparison of the amounts of clothesline needed to make two different types of clothes-drying rack. In the first rack the pieces of clothesline are strung in parallel between two supports, while in the second rack the pieces of clothesline are strung in the form of squares of different sizes between two cross-bars. According to Kilpatrick (1987), such a problem gives the opportunity to develop different mathematical models depending on the amount of real-world knowledge and realistic considerations that one allows in the formulation and solution of the problem. For example, is it necessary to take into account the clothesline that would be needed to tie the different pieces of rope to the supports so that they would actually stay up? Could we get along without all the separate lengths (which entail a lot of measuring, cutting and knotting) and instead solve the problem using one single length of clothesline for each pattern? After the students have explored this problem in small groups, the class could make an overview of the different mathematical models that were used, and

Tom wants to make a clothes-drying rack for the backyard. He can fix it so that the clotheslines are strung between two supports as in Figure (a) or between cross bars as in Figure (b).

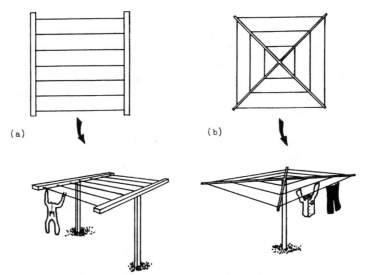

How many feet of clothesline would he have to get for each of these choices if the outer square measures 6 feet on a side and the separation between adjacent lines is 1 foot?

Fig. 1.2   Sample problem proposed by Kilpatrick (1987, p. 126) as a good basis for a classroom discussion on mathematical modeling. Copyright 1987 by Lawrence Erlbaum Associates. Reprinted with permission.

discuss the adequacy, the precision, etc. of the different models. Another theme of the class discussion could be that the problem itself is reformulated in different ways leading to different mathematical models (Kilpatrick, 1987).

Summary and discussion

The observations and studies reviewed in this chapter vary in terms of the nature of the problems used as well as the research questions the authors were concerned with and the theoretical frameworks wherein these questions were raised. In particular, the questions such as "How old is the captain?" are nonsensical, whereas questions such as "How much to post a letter?" have a definite real-world answer; questions such as asking what height a man would be when 20 years old given his height when 10 years at least admit of a reasonable estimation. Nevertheless, all the studies have certain characteristics in common. They all show that students being confronted with word problems in a typical school setting are engaged in a peculiar kind of activity wherein they typically solve these problems in a stereotyped and artificial way without relating them to any real-life experience. One of the major rationales underlying the use of word problems in school is the assumption that these texts are condensed but genuine descriptions of real life situations and – consequently – that students' word-problem-solving activities appropriately mirror those of people applying mathematics in real-world settings. However, the above examples illustrate that it means a different thing to have a mathematics problem in school and in an out-of-school setting, and that, therefore, there is no guarantee that students' experiences with school word problems will "travel serviceably across the school-life bridge" (Lave, 1992, p. 79) into the everyday world (see also De Corte et al., 1996; Nunes et al., 1993; Resnick, 1987; Saxe, 1988). Rather than functioning as realistic contexts that invite, or even force, students to use their common-sense knowledge and experience about the real world in combination with acquired mathematical knowledge and skills, school arithmetic word problems seem to be perceived by students as a capricious kind of school tasks that are separated from the real world and that have to be solved by means of certain computational techniques on the given numerical data, ignoring real-world knowledge and even accepting conditions about the problem context that are empirically false.

In contrast to the model of competent mathematical modeling presented in Figure 1.1, we assert that the process of solving word problems for many students is often along the lines of the model presented in Figure 1.3, wherein several steps of the modeling process are more or less completely bypassed. According to this model, the problem text immediately guides the choice of one (or more) of the four arithmetic operations – a choice that may be based on superficial features such as the presence of certain key words in the text (for instance, the word "less" in the problem text automatically results in a decision to perform a subtraction) or on an association between the situation described

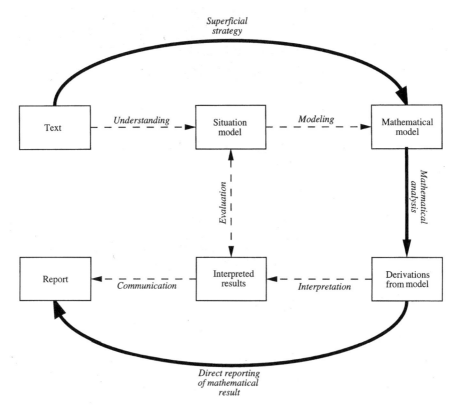

Fig. 1.3   Superficial solution of a word problem.

in the problem text and a primitive model for one of the operations (for instance, anything that suggests the act of "putting together", triggers the operation of addition). The directly evoked operation is then applied to the numbers embedded in the problem text and the result of the calculation is found and given as the answer, typically without reference back to the problem text to verify that it is a meaningful answer to the original question and/or to check for reasonableness.

According to many authors, such a superficial and artificial solution process, wherein both the processes of understanding and modeling and those of interpreting and evaluating are bypassed, is promoted and reinforced by the stereotyped nature of word problems as typically presented and by the kind of talk and activity around these problems in the traditional mathematics classroom (De Corte & Verschaffel, 1989; Nesher, 1980; Reusser, 1988; Schoenfeld, 1991).

The illustrative "disaster stories" described in this first chapter stimulated our interest in the topic and led to the two exploratory studies that are reported in the next chapter. While our interest was aroused, we were not aware at the outset of the depth of the theoretical issues raised by these examples and of the methodological complexities involved in studying them in a more systematic way.

2

Suspension of Sense-making: Ascertaining the Problem

Question: John's best time to run 100 meters is 17 seconds. How long will it take him to run 1 kilometer?
Answer: 10 × 17 = 170 seconds.
(Verschaffel et al., 1994)

The extent to which students apparently ignore plausibly relevant and familiar aspects of reality in answering word problems was simultaneously studied in a systematic way among 13–14-year-olds in Northern Ireland (Greer, 1993) and 10–11-year-olds in Belgium by Verschaffel et al. (1994). We begin by summarizing the design and results of these two studies. The two initial studies were replicated with a high degree of consistency in several countries, namely Belgium, Germany, Japan, Northern Ireland, Switzerland, and Venezuela and these replication studies are (partly) discussed in the next section. In some of these replications, paper-and-pencil tests were complemented by individual interviews with a small number of pupils, and the results and interpretation of these are reported next. Taken as a whole, these studies confirm the overall conclusion from the "disaster stories" reviewed in Chapter 1, documenting the prevailing situation whereby students answer school word problems with apparent scarce regard for whether the answers make sense when considered from the viewpoint of the real-world situations verbally described in those problems.

Unrealistic responses to word problems: Northern Ireland study

A total of 100 13- and 14-year-olds from two classes from each of two schools in Northern Ireland was tested by Greer (1993). Items were constructed in eight pairs (Table 2.1), six of which pairs comprise one item that can be solved straightforwardly by applying the most obvious arithmetic operation with the numbers in the problem, yoked with a more problematic item for which the simple arithmetical operation offers an inappropriate, or at best weak, model if one seriously takes into account the realities of the context evoked by the problem statement. Some of the items were taken directly, or adapted, from the literature (such as the problem about the buses (see pp. 6–8) and the problem about the cost of posting a letter (see pp. 8–9)), while others were constructed by the author. The first three pairs relate to various aspects of division, the others to proportionality. Each student responded to one item from each pair, and the items for each student were equally divided between straightforward and problematic versions. The instructions given were minimal and non-directive, but included an invitation to comment on the question or the answer. Students' answers were categorized on an ad hoc basis.

As expected, virtually no errors were made on the straightforward items. For the problematic items, on the other hand, a large majority of responses showed no adjustment for realistic constraints, either in terms of the answer given, or in terms of a comment on the answer. For the problems about the pieces of rope being tied together (Pair 2), the time to run 3 miles (Pair 4), the filling of the flask of diminishing cross-section (Pair 5), and the number of animal names beginning with C generated in 3 minutes (Pair 6), a very large majority of students responded as if they were no different in essence from the associated straightforward items. Thus, 83% of the students answered the rope problem with "8 pieces" without any comment; 90% responded as if direct proportionality were appropriate for the athlete's running time; on the flask problem, also, 90% of the students' responses were in accordance with direct proportionality (without any comment); and the item about the animal names elicited 94% responses based on direct proportional reasoning without any further comment. On the other hand, 83% of the students recognized that each child should get 3 rather than 3.5 balloons (Pair 1), 75% gave non-proportional answers and 24% gave proportional answers accompanied with a comment or explanation acknowledging that the sale of Christmas cards is not uniform across the year or that the proportionate answer would be an estimate (Pair 7). The busing and similar eggs problems (Pair 3) elicited a considerable number of realistic reactions too (61% and 58%, respectively). Finally, only 6% overall of the students showed indications of assuming linearity for the mail costs problem (Pair 8). This latter finding is discrepant from the findings of Säljö and Wyndhamn (1990) and Bender (1985), discussed in the previous chapter (pp. 8–12), who found much higher percentages of children producing unrealistic answers based on direct proportionality.

In sum, while there were striking differences between the eight problematic items, the results confirmed Greer's (1993) general hypothesis that students

Table 2.1   Problems Used in Greer's (1993) Study.*

Pair 1
If there are 14 pizzas for 4 children at a party, how should they be shared out?
If there are 14 balloons for 4 children at a party, how should they be shared out?

Pair 2
A man cuts a piece of rope 12 meters long into pieces 1.5 meters long. How many pieces does he get?
A man wants to have a rope long enough to stretch between two poles 12 meters apart, but he only has pieces of rope 1.5 meters long. How many of these would he need to tie together to stretch between the poles?

*Pair 3***
1128 eggs are being packed into boxes. Each box can hold 36 eggs. How many boxes will be needed?
1128 children are going on a trip in buses. Each bus can carry 36 children. How many buses will be needed?

Pair 4
A barge travels a mile in 4 minutes and 7 seconds. About how long would it take to travel 3 miles?
An athlete's best time to run a mile is 4 minutes and 7 seconds. About how long would it take him to run 3 miles?

Pair 5
The flask is being filled from a tap at a constant rate. If the depth of the water is 2.4 cm after 10 seconds, about how deep will it be after 30 seconds?

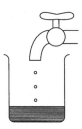

The flask is being filled from a tap at a constant rate. If the depth of the water is 2.4 cm after 10 seconds, about how deep will it be after 30 seconds?

Table 2.1 continues

Table 2.1 (continued)

Pair 6
A girl is counting cars going past her house. In one minute she counts 9 cars. About how many will she count in the next 3 minutes?
A girl is writing down names of animals that begin with the letter C. In one minute she writes down 9 names. About how many will she write in the next 3 minutes?

Pair 7
A shop sells 312 birthday cards in December. About how many do you think it will sell altogether in January, February, and March?
A shop sells 312 Christmas cards in December. About how many do you think it will sell altogether in January, February, and March?

Pair 8
John's letter costs 64p to send first class. Mary's letter weighs exactly twice as much. How much does it cost to send first class? (Accompanied by the then current Post Office table with first class prices of different weights)
John's letter costs 64p to send first class. Mary's letter weighs exactly twice as much. How much does it cost to send first class? (Accompanied by the then current Post Office table with first class prices for different weights, with an extra cue potentially evoking direct proportional reasoning)

---

*   Where applicable, the straightforward item in the pair comes first, and the complex one second.
**   In this case there are two complex items, both requiring 32 as the answer.

are liable to respond to word problems according to stereotyped procedures assuming that the modeling of the situation described is "clean". In other words, for many of the problems students produced answers without apparently considering each textually represented situation on its merits, but rather by uncritically carrying out an arithmetic operation with the numbers in the problem.

By way of suggestive illustration of how the pupils regarded the items, two responses to the Christmas cards item (Pair 7) were particularly interesting. A number of students hedged by stating that the number of cards sold in January to March would be small, but hedged by giving the proportionate answer as what would happen if sales were constant. One pupil carried out a proportional calculation, not by multiplying by 3, but taking into account the different numbers of days in the months, and so multiplying by 90/31 (assuming not a leap year).

Unrealistic responses to word problems: Flemish study

The subjects in Verschaffel et al.'s study (1994) were 75 pupils (10–11-year-old boys and girls) from three fifth-grade classes of three Flemish schools in which word problem solving was taught in the traditional way. This implies that the pupils had frequently been confronted with traditional school word problems rather than with authentic problem situations, and that realistic modeling had not been systematically addressed in teaching (for a description of this traditional teaching practice, see De Corte & Verschaffel, 1989).

A paper-and-pencil test was constructed consisting of ten matched pairs of items (Table 2.2). As for most of the pairs in Greer's (1993) study, each pair consisted of:

- A standard item (S-item) that can be solved unproblematically by applying the most obvious arithmetic operation(s) with the given numbers (e.g., "Steve has bought 5 planks of 2 meters each. How many planks of 1 meter can he saw out of these planks?").
- A parallel problematic item (P-item) for which the appropriate mathematical model is less obvious, at least if one seriously takes into account

Table 2.2   Ten Item Pairs Involved in Verschaffel et al.'s (1994) Study.*

Pair 1
S1   Pete organized a birthday party for his tenth birthday. He invited 8 boy friends and 4 girl friends. How many friends did Pete invite for his birthday party?
P1   Carl has 5 friends and Georges has 6 friends. Carl and Georges decide to give a party together. They invite all their friends. All friends are present. How many friends are there at the party? *(Friends item**)*

Pair 2
S2   Steve has bought 5 planks each 2 meters long. How many planks 1 meter long can he saw from these planks?
P2   Steve has bought 4 planks each 2.5 meters long. How many planks 1 meter long can he saw from these planks? *(Planks item)*

Pair 3
S3   A shopkeeper has two containers for apples. The first container contains 60 apples and the other 90 apples. He puts all the apples into a new, bigger container. How many apples are there in that new container?
P3   What will be the temperature of water in a container if you pour 1 liter of water at 80° and 1 liter of water at 40° into it? *(Water item)*

Pair 4
S4   Pete's piggy bank contains 690 francs. He spends all that money to buy 20 equally priced toy cars. How much was the price of one toy car?
P4   450 soldiers must be bussed to the their training site. Each army bus can hold 36 soldiers. How many buses are needed? *(Buses item)*

Pair 5
S5   A boat sails at a speed of 45 kilometers per hour. How long does it take this boat to sail 180 kilometers?
P5   John's best time to run 100 meters is 17 seconds. How long will it take him to run 1 kilometer? *(Runner item)*

Pair 6
S6   Chris made a walking tour. In the morning he walked 8 kilometers and in the afternoon he walked 15 kilometers. How many kilometers did Chris walk?

Table 2.2 continues

Table 2.2 (continued)

P6 Bruce and Alice go to the same school. Bruce lives at a distance of 17 kilometers from the school and Alice at 8 kilometers. How far do Bruce and Alice live from each other? *(School item)*

Pair 7
S7 Kathy, Ingrid, Hans and Tom got from their grandfather a box with 14 chocolate bars, which they shared equally amongst themselves. How many chocolate bars did each grandchild get?
P7 Grandfather gives his 4 grandchildren a box containing 18 balloons, which they share equally. How many balloons does each grandchild get? *(Balloons item)*

Pair 8
S8 This morning Steve had 1480 francs in his piggy bank. Now he has 1650 francs in his piggy bank. How many francs did Steve gain since this morning?
P8 Rob was born in 1978. Now it's 1993. How old is he? *(Age item)*

Pair 9
S9 A man cuts a clothesline of 12 meters into pieces of 1.5 meters each. How many pieces does he get?
P9 A man wants to have a rope long enough to stretch between two poles 12 meters apart, but he has only pieces of rope 1.5 meters long. How many of these pieces would he need to tie together to stretch between the poles? *(Rope item)*

Pair 10
S10 This flask is being filled from a tap at a constant rate. If the depth of the water is 4 cm after 10 seconds, how deep will it be after 30 seconds?

P10 This flask is being filled from a tap at a constant rate. If the depth of the water is 4 cm after 10 seconds, how deep will it be after 30 seconds? *(Flask item)*

---

*   The symbol S refers to "standard", and the symbol P to "problematic".
**   Each P-item has been given a short verbal label for identification.

the realities of the context evoked by the problem statement (e.g., "Steve has bought 4 planks of 2.5 meters each. How many planks of 1 meter can he saw out of these planks?").

The ten pairs of problems were administered to each pupil in two series on the same day. Each series contained the P-version of five problem pairs and the S-version of the five other pairs. To control for ordering effects, the problems in each series were presented in two different orders, and in each class one half of the children started with one series while the other half was given the other series first. The administration of the problems was done by the class teacher as part of a normal mathematics lesson. With respect to each problem, pupils were asked to write down not only their answer, but also how they arrived at this answer (e.g., by recording the calculations) and possible other comments they might have.

As in Greer's (1993) study, children's reactions to the problems were analyzed in two ways for evidence of the activation and use of real-world knowledge and realistic considerations about the problem context, namely by (a) distinguishing in their answers and computations between realistic answers and non-realistic ones, and (b) looking for evidence of awareness of realistic constrains in their additional comments.

When a pupil either gave an answer to the problem that was scored as realistic or produced a non-realistic answer that was accompanied by a realistic comment, his (her) overall reaction to that particular problem was scored as a "realistic reaction" (RR). Take, for example, the planks item mentioned above. An RR categorization was given not only to a child who produced the (most) realistic numerical answer "$4 \times 2 = 8$; he can saw 8 planks of 1 meter", but also to a child who responded with "10 planks" but added the comment that "Steve would have a hard time putting together the remaining pieces of 0.5 meters". The categorization NR ("non-realistic reaction") was given to children who answered the problem in a non-realistic manner and did not give any further realistic comment (e.g., answering the planks problem as follows: "$4 \times 2.5 = 10$; he can saw 10 planks of 1 meter"). Table 2.3 contains typical examples of answers considered as NR and as RR for each of the ten P-items (for more details about the scoring system, see Verschaffel et al., 1994).

The overall hypothesis of the first study was that due to their extensive experience with an impoverished diet of standard word problems, and to the lack of systematic attention to the mathematical modeling perspective in their mathematics lessons, pupils would demonstrate a strong tendency to exclude real-world knowledge and contextual considerations from their problem-solving endeavors on the P-items, and consequently would solve them as if they were not at all problematic.

Table 2.4 gives the number of pupils who reacted in a realistic (RR) and in a non-realistic (NR) way for each of the ten P-items from Table 2.2.

The data in Table 2.4 strongly support the hypothesis. As predicted, the pupils demonstrated a very strong overall tendency to exclude real-world

Table 2.3 Typical Examples of Non-Realistic (NRs) and Realistic Reactions (RRs) to the Ten P-Problems (Verschaffel et al., 1994).

Friends	NR: $6 + 5 = 11$. There will be 11 friends at the party. RR: You cannot know how many friends there will be at the party
Planks	NR: 4×2.5 meters $= 10$ meters. 10 meters $\div$ 1 meter $= 10$. Steve can saw 10 planks of 1 meter. RR: Steve can saw 2 planks of 1 meter from one plank of 2.5 meters. $2 \times 4 = 8$. So, Steve can saw 8 planks.
Water	NR: $80° + 40° = 120°$. RR1: $80° + 40° = 120°$; $120° \div 2 = 60°$. RR2: I don't know exactly. It must be something in between 80° and 40°.
Buses	NR1: 450 divided by 36 is 12.5, so 12.5 buses are needed. NR2: 450 divided by 36 is 12.5, so 12 buses are needed. RR: 450 divided by 36 is 12.5, so 13 buses are needed.
Runner	NR: $17 \times 10 = 170$. John's best time to run 1 kilometer is 170 seconds. RR1: It is impossible to answer precisely what John's best time on 1 kilometer will be. RR2: About three and a half minutes. RR3: Certainly more than 170 seconds.
School	NR1: $17 - 8 = 9$. Saskia and Bruno live 9 kilometers from each other. NR2: $17 + 8 = 25$. Saskia and Bruno live 25 kilometers from each other. RR1: You cannot know how far Saskia and Bruno live from each other. RR2: The answer must lie between 9 and 25.
Balloons	NR: $18 \div 4 = 4.5$ balloons for each child. RR: $18 \div 4 = 4.5$, so there will be 4 balloons for each child (with two balloons left over).
Age	NR: $1978 + 15 = 1993$. He is aged 15. RR1: 14 or 15. RR2: Cannot be known precisely.
Rope	NR: $12 \div 1.5 = 8$; so 8 pieces of 1.5 meters are needed. RR1: It is impossible to know how many pieces of rope you will need. RR2: Certainly more than 8 pieces.
Flask	NR: $3 \times 4 = 12$. After 30 seconds, the level of the water will be 12 cm. RR1: It is impossible to give a precise answer. RR2: More than 12 cm (because of the shape of the flask). RR3: I don't know exactly. Maybe 14 or 15.

knowledge and realistic considerations when confronted with the problematic versions of the problems. In total, only 128 out of the 750 reactions to the P-items (17%) could be considered as realistic (RR), either because the pupil wrote a realistic answer or made an additional realistic comment. For only two out of the ten P-items was a considerable number of realistic answers or comments observed: the buses item (P4) and the balloons item (P7) which elicited 49% and 59% RRs, respectively.

Table 2.4   Percentages of Realistic Reactions (RRs) on P-items in the Study of
Verschaffel et al. (1994).*

Problem	% of RRs
Friends	11
Planks	14
Water	17
Buses	49
Runner	3
School	5
Balloons	59
Age	3
Rope	0
Flask	4
Total	17

*   For the full text of the P-items listed in the first column, see Table 2.2. For a detailed
description of the criteria for scoring a reaction as NR or RR, see Verschaffel et al.
(1994).

Replications of the two pioneering studies

The findings of Greer (1993) and Verschaffel et al. (1994) have been replicated in several countries (namely Belgium, Germany, Japan, Northern Ireland,
Switzerland, and Venezuela), sometimes as part of more extensive studies
about the effects of certain variations in the presentation of the P-items or in
the testing setting. Because of these alterations in the testing conditions of
these studies, we will return to them in the next chapters (Chapters 3 and 4).
However, to demonstrate the universality and the consistency of the findings
observed in the original two studies, we report here partial results from five
other studies in which, sometimes as a first step or as a part of a larger investigation, groups of students were administered some or all of the ten P-items
used by Verschaffel et al. (1994) under the same testing conditions and using
the same procedure for data analysis as in Greer's (1993) and/or Verschaffel
et al.'s (1994) original studies. More specifically, these data are from the
following six studies:

- A study by Caldwell (1995), in which five P-items from the test of
  Verschaffel et al. (1994), together with two items from Greer's (1993) problem set, were administered to 75 10–11-year-old boys and girls from three
  different seventh grade classes in three different schools in Northern Ireland.
- A study by Hidalgo (1997), in which Verschaffel et al.'s (1994) test was
  taken by 119 fifth-graders from four different schools in Venezuela.
- A study by Reusser and Stebler (1997a), in which the same test was administered to 439 secondary school students aged 13, randomly selected from
  seventh-grade classes across the German-speaking part of Switzerland.

- A study by Yoshida, Verschaffel, and De Corte (1997), who administered the same test to 91 fifth-graders from three randomly selected classes in a Japanese primary school. The data reported here concern only half of these pupils, namely those who received the test under the same experimental conditions as in Verschaffel et al.'s (1994) study; the results for the other half are reported in Chapter 3.
- A study by Renkl (1999), in which eight out of the ten P-items from the test of Verschaffel et al. (1994) were given to 93 German fourth-graders at the end of the school year. Three of these P-items, namely the friends item (P1), the buses item (P4), and the balloons item (P7) were slightly reformulated by Renkl to make the problematic nature of these P-items even clearer than in Verschaffel et al.'s (1994) original study.
- A study by Verschaffel, De Corte, and Lasure (1999), in which seven of the P-items used in Verschaffel et al.'s (1994) first study, together with their corresponding S-items, were collectively administered to a group of 64 fifth-graders from three different schools in Flanders, as the first part of a follow-up study that is reported in full in Chapter 3.

As shown in Table 2.5, the results of all these studies were very much in accordance with those obtained by Greer (1993) and Verschaffel et al. (1994). The vast majority of students demonstrated little or no tendency to include real-world knowledge into the solving of most of the P-items, and the relative percentages of RRs for the distinct P-items were very similar to those in the original studies. A particularly noticeable recurrent finding is that the two problems about division with a remainder (i.e., the buses and the balloons items) elicited always many more RRs than the other P-items.

Interviews with students about their responses

The outcomes of the investigations described so far provide empirical evidence for pupils' strong tendency to exclude relevant real-world knowledge and realistic considerations from their understanding and solving of school arithmetic word problems. However, these findings need to be put into perspective because of some methodological limitations of the testing procedure used in these studies. In this section two studies will be reported, one by Caldwell (1995) and the other by Hidalgo (1997), in which a serious effort was made to overcome these methodological weaknesses.

A first methodological limitation is the lack of empirical data about students' mastery of the real-world knowledge that is necessary to respond to or comment on the different P-problems in a realistic way. While it seems reasonable to assume that this knowledge is available to a typical upper elementary pupil (most 10-year-olds will probably know that there exist no half-buses or that making a knot in a rope will shorten its total length) this might not always be the case. Consequently, in some cases, students' failure to give a realistic

Table 2.5   Percentages of Realistic Reactions (RRs) on P-items in the Studies of Greer (1993) (G), Verschaffel et al. (1994) (V1), Caldwell (1995) (C), Hidalgo (1997) (H), Reusser and Stebler (1997a) (RS), Yoshida et al. (1997) (Y), Renkl (1999) (R), and Verschaffel, De Corte, and Lasure (1999) (V2).*

Problem	G	V1	C	H	RS	Y	R	V2
Friends	–	11	5	23	11	13	10	14
Planks	–	14	–	16	14	0	21	17
Water	–	17	–	11	21	11	9	–
Buses	55	49	65	11	49	62	67	64
Runner	6	3	0	0	5	7	2	1
School	–	3	–	1	5	2	1	9
Balloons	85	59	81	55	75	52	51	–
Age	–	3	–	0	2	0	–	–
Rope	8	0	1	0	6	2	0	5
Flask	2	4	–	0	0	4	–	5
Total**	–	17	–	12	19	15	–	–

*   For the full text of the P-items listed in the first column, see Table 2.2. For a detailed description of the criteria for scoring a reaction as NR or RR, see Verschaffel et al. (1994).
**   Mean percentages of RRs are shown only for those studies in which all 10 P-items were administered.

reaction may have been caused by the *absence* of the knowledge of (or by misconceptions about) the contextual elements involved in the problem, rather than by *failure to apply* it. For example, some children may actually believe that putting together two hot containers of water increases the heat (P3 problem in Table 2.2), or may have imperfect understanding of the factors that determine the rise of the level of liquid in a container (P10 problem in Table 2.2). Commenting on the disastrous results on the absurd problems about the captain's and the shepherd's ages obtained by the IREM of Grenoble (Chapter 1, pp. 3–6), Freudenthal (1982) pointed to lack of understanding of the context involved (i.e., of what the age of a person is) as a possible explanation. Neither in the studies by Greer (1993) and Verschaffel et al. (1994), nor in any of the replication studies did the researchers systematically control for this possible confounding factor.

A second important restriction of the studies reported so far derives from the technique used to gather the data, namely a collective paper-and-pencil test. The problem with this technique is that it does not yield direct information about the reasons for students' apparent tendency to pay hardly any attention to the reality constraints of most of the presented P-items (assuming that they possess the relevant knowledge). For instance, one could argue that some students might have activated real-world knowledge during their solution processes for P-items *which was not reflected in their written protocols*, simply because they finally decided to respond and react in a "conformist" rather than a "realistic" way, in line with their beliefs and conceptions about "the rules of the game" of school arithmetic word problems (De Corte & Verschaffel, 1985; see also Chapter 5). This tendency may have contributed to an underestimation of the number of solution processes in which there was activation of relevant real-world knowledge by the students. More detailed and more process-oriented information about students' beliefs and conceptions underlying their tendency to exclude real-world knowledge from their word problem solving endeavors can be obtained by means of individual interviews as a complementary data-gathering technique.

In the studies of Caldwell (1995) and of Hidalgo (1997), certain pupils who had solved the written test were afterwards selected for participation in individual interviews aimed at unraveling the reasons for the apparent neglect of real-world knowledge and context-based considerations in their solutions of (most of) the P-items. Because in these interviews special attention was also paid to the issue of the mastery of the relevant real-world knowledge, we can consider these two investigations as attempts to overcome the two above-mentioned restrictions of the studies reviewed so far.

In Caldwell's study (1995) a small number of children were individually asked to explain verbally how they had derived their (non-realistic) answers to the P-items during the paper-and-pencil test. Although the data from these interviews are not reported and discussed in a systematic and detailed way, they nevertheless illustrate that at least some pupils were able to articulate an awareness of a difference between conventional answers expected in the context of school mathematics, on the one hand, and answers appropriate to real situations, on the other. For instance, one 10-year-old commented as follows in response to the interviewer's question as to why she did not make use of realistic considerations when solving the P-items in the context of the written test (Caldwell, 1995, p. 39): "I know all these things, but I would never think to include them in a maths problem. Maths isn't about things like that. It's about getting sums right and you don't need to know outside things to get sums right." The interviews suggested also that relevant information (such as that the water will rise more quickly as the flask narrows, or that a runner tires with distance) was indeed generally known to these children. These findings weaken the alternative interpretation of the alarmingly weak results on the P-items mentioned above, namely that students' failure to give an RR to a P-item might be caused by the absence of knowledge about (or by a misconception about) the context involved in the problem, rather than by not applying that knowledge.

In Hidalgo's study (1997) a group of 15 pupils who had produced a considerable number of NRs during the paper-and-pencil test were also selected to participate in individual interviews. During the interview, the pupil was confronted again with some P-items. For every item administered in the interview, (s)he was asked to read it aloud, to explain its meaning, and to solve it while thinking aloud. Finally, the pupil was explicitly instructed to check the correctness of his/her solution. By asking specific questions, Hidalgo also tried to assess whether the pupils possessed the relevant real-world knowledge they did not use when solving the P-items during the paper-and-pencil test. Based on the analysis of these interview data, Hidalgo (1997) tried to evaluate the relevance of the following hypothetical reasons for students' NRs on the P-items:

- Beliefs about school arithmetic word problems, and, more specifically, the belief that every word problem has to be solved by a single, numerical answer which is the result of one or more operations with the numbers given in the problem.
- Unfamiliarity with the particular context involved in the problem (e.g., mixture of liquids with different temperatures, best running times, etc.) and lack of necessary real-world knowledge to build an adequate and realistic representation of the problem situation.
- Lack of important heuristic and metacognitive skills, such as carefully reading and analyzing the problem situation and checking whether the outcome of the computational work makes sense in relation to the original problem situation.

Based on a detailed qualitative analysis of the pupils' reactions to the different tasks of the interview, Hidalgo (1997) concluded that a high proportion of the NRs was the result of a combination of the three hypothetical explanatory factors listed above, but that the relevance of each factor differed from subject to subject and from problem to problem. With respect to the first explanatory factor, several thinking-aloud protocols contained clear articulations of pupils' (mis)beliefs about what a word problem is and how it should be solved in a school context (similar to the articulated belief of a pupil from Caldwell's study quoted above). With respect to the second factor, she did find – in contrast to Caldwell's (1995) second finding – that some pupils failed because of their unfamiliarity with or deficient knowledge of the phenomenon described in the problem (especially for the flask problem). Third, the analysis of the solution processes revealed that many pupils' solution processes did not involve an analysis phase or metacognitive control. In line with what is found in many other studies (De Corte et al., 1996; Kilpatrick, 1985; Schoenfeld, 1992; Verschaffel, 1999), many pupils' solution strategy consisted of quickly reading the problem and hastily deciding upon what calculation to perform, without any serious problem analysis before or any serious interpretation and evaluation after the actual calculational activity, as in the model shown in Figure 1.3. If pupils' word-problem-solving activities are restricted to the selection and execution of one or more computations with the numbers given in the problem statement (neglecting the other stages of competent problem solving such as problem analysis and metacognitive control), there is, of course, little chance that they will be alerted to realistic constraints of the problem context that would call into question the correctness of their proposed routine solutions.

In sum, the interview data from the studies of Caldwell (1995) and Hidalgo (1997) already yield some empirical evidence about the cognitive processes that underlie the large number of NRs on the P-items in the paper-and-pencil studies reported above. Taken as a whole, they suggest that, while unfamiliarity with the contexts involved in the problems and lack of appropriate heuristic and metacognitive skills may provide contributory explanations, it is the more or less conscious (mis)beliefs about school arithmetic word problems that constitute the major reason why so many pupils solve the P-items in a non-realistic way.

Summary and discussion

The goal of the studies of Greer (1993) and Verschaffel et al. (1994) was to collect in a systematic way empirical data about the (lack of) activation of real-world knowledge during students' understanding and solution of arithmetic word problems in a school context. Therefore they constructed a set of problems, partly relying on examples they had encountered in the literature (see Chapter 1), in which the relationship between the context described in the problem statement and the putatively corresponding mathematical operation(s) was not simple and straightforward, at least if one seriously takes into account the realities of that particular context. The analysis of students' reactions to these P-items in both studies, as well as in a number of replication studies executed in different parts of the world, provided further demonstration of students' strong tendency to exclude real-world knowledge and realistic considerations from their solution of these problems. Only two of the P-items used in these studies elicited a considerable number of responses reflecting adjustments for realistic constraints or qualifications based on realistic considerations, namely the two DWR problems, one of which required rounding up to the next whole number (the buses item) and one of which required rounding down to the lower whole number (the balloons item). For all other P-items, the number of students who made such realistic adjustments of qualifications was very low.

Two studies in which collective paper-and-pencil tests were complemented with individual interviews provided some further insight into the reasons for students' tendency to pay hardly any attention to even the simplest reality constraints of most of the presented P-items. Taken together, they suggest that, while unfamiliarity with the context of the problems and lack of appropriate skills for problem analysis and evaluation may be complementary causal factors, the main reason why so many pupils solved the P-items in a non-realistic way is to be found in their (mis-)beliefs about school arithmetic word problems. However, the documentation by Hidalgo (1997) of the absence of real-world knowledge as a possible reason for at least some of the NRs implies a methodological weakness in the studies of Greer (1993) and Verschaffel et al. (1994), wherein the availability of the necessary real-world knowledge was intuitively assumed rather than empirically guaranteed or checked.

Although none of the studies reported in this chapter directly addresses the question as to what instructional factors are responsible for the development of the tendency toward non-realistic mathematical modeling among students, this undesirable learning outcome is considered by most authors to be the result of the interplay between (1) the stereotyped and straightforward nature of the vast majority of the word problems students encounter in their mathematics lessons, and (2) the nature of the teaching and learning activities taking place with these problems. From these aspects of current classroom practice and culture, it seems clear that students slowly but surely construct the belief that making realistic considerations and elaborations about the situation described in a school arithmetic word problem does more harm than good in meeting the requirements of the instructional situation. In Chapter 5, we review and discuss theoretical analyses of these instructional factors as well as the empirical evidence supporting the claim that these factors are indeed responsible for the origin and the development of this belief among pupils. We will also argue that this belief cannot be seen in isolation from other aspects of the "didactical contract" (Brousseau, 1984, 1990, 1997) that direct in an implicit but almost irresistible way the thinking processes and the performances of students when being instructed in or tested on mathematical applications in a traditional school setting.

Before getting to that discussion, two lines of follow-up studies will be reviewed in Chapters 3 and 4. Whereas the first line of research tests the minimal hypothesis that the extremely low percentages of RRs on the P-items in the studies reported in this chapter can be simply explained in terms of the students' lack of attention or misleading test instructions, the second line of research looks at the effectiveness of more drastic changes in the experimental setting, namely attempts to increase the authenticity of the testing setting. The importance of these follow-up studies is twofold. On the one hand, they yield further insight into the relevance of certain hypothetical explanations of students' tendency to pay hardly any attention to even the simplest reality constraints on most of the P-items. On the other hand, because they involve minimal interventions on the part of the investigators, they may be considered as a first step beyond the mere documentation of the state of affairs with respect to word problems existing in mathematics classrooms.

3

Effects of Alerting Students about Problematic Word Problems

> *I did think about the difficulty, but then just calculated it the usual way.*
> *(Why?) Because I just had to find some sort of solution to the problem, and that was the only way it worked. I've got to have a solution, haven't I?*
> *(Response of a student to the researcher's question as to why he answered the friends and the runner item in the usual non-realistic way (Reusser & Stebler, 1997a, p. 317))*

In this chapter we discuss the design and results of a first series of studies which constitute an initial step beyond the mere documentation of the state of affairs with respect to students' tendency towards non-realistic modeling of school arithmetic word problems. Common to these studies is that they involve minimal interventions taking the form of giving students a hint that some of the problems need more careful consideration and/or giving them direct and explicit help to consider alternative responses taking into account realistic considerations. The answers of these students are compared to those who received the same P-items under the original, "neutral" testing conditions.

The common rationale behind all of these studies can be summarized as follows. Both Greer (1993) and Verschaffel et al. (1994) interpreted their striking results as evidence for the basic hypothesis that by the end of the elementary school or the beginning of the secondary school the vast majority of students have learned to exclude realistic considerations from their interpretations and solutions of school arithmetic word problems. However, one could argue that the fact that so many students in these studies reacted in such an apparently mindless and unrealistic way was merely an artifact of the experimental setting. More specifically, the formulation of the P-items (i.e., the fact that the P-items were formulated similarly to the S-items), the way in which these problems were presented to the students (mixed with S-items), and the accompanying instructions (e.g., the lack of any warning that this was not a standard test, the absence of an explicit invitation to provide alternative answers or even to criticize the problems) all may have contributed to the expectation among the students that this was a test involving only problems that had to be conceived, handled and solved in the standard manner. Other task formats and administration conditions might have produced completely different, and hence much less striking, results. Moreover, the technique of paper-and-pencil tests used in these studies may have concealed the fact that, even under these unfavorable testing conditions, some students might have effectively thought about the problematic or unsolvable nature of (at least some of) the P-items, but finally decided not to incorporate these considerations into their final (written) responses, because they assumed that they had to approach the problems in the "usual" way. Greer (1997, p. 305) has introduced the term "experimental contract", by analogy with the concept of "didactical contract", to stress the importance of giving more consideration to how students acting as subjects construe what is happening when being tested or interviewed, and how this construal influences their responses.

For all these reasons, it seemed premature to take the results from the studies of Greer (1993) and Verschaffel et al. (1994) as decisive empirical evidence for the conjecture that several years of exposure to the current instructional practice of word problem solving have shaped these pupils to produce mindless and unrealistic responses to word problems. If one could succeed in drastically decreasing the number of NRs to the P-items by means of some simple variations in the way the P-items are phrased, presented or introduced to the students, there would be much less need to be concerned, or even shocked, about the results obtained in these initial studies. Five studies involving such variations in the task format and the administration conditions are reported in this chapter.

Effect of general hints: A Japanese study

Besides the replication of the original study by Verschaffel et al. (1994) with Japanese pupils, reported in Chapter 2 (pp. 23–24), Yoshida et al. (1997) also made a comparison between groups of Japanese pupils with and without extra hints aimed at improving the disposition towards more realistic mathematical problem solving. The 91 fifth-graders involved in that study were randomly assigned to one of two conditions: 45 fifth graders were assigned to condition I (neutral condition) and the remaining 46 pupils were assigned to condition II (warning condition). While in condition I the test was administered basically in the same way as in the study of Verschaffel et al. (1994), pupils in condition II were given the following extra written instruction at the top of their test sheet: "The test contains several problems that are difficult or impossible to solve because of certain unclarities or complexities in the problem statement. When you encounter such an unclarity or complexity, please write it down and explain why you think that you are not able to solve the problem". The procedure for analyzing the data was exactly the same as in Verschaffel et al. (1994). The percentages of RRs were 15% and 20% for the pupils in conditions I and II, respectively. A t-test comparing the results of these two groups of Japanese pupils revealed that the difference in the total scores of these groups failed to reach significance at the 5% level. So, the additional general instruction at the start of the test, aimed at increasing the alertness of the pupils and thereby the number of RRs on the P-items, produced only a small, statistically non-significant difference in favor of the group receiving that instruction. This lack of a strong effect attributable to the extra hint suggests that pupils' tendency to exclude realistic considerations from their interpretation of arithmetic word problems in a typical school setting is deeply entrenched and resistant to change.

Effect of varying test presentation: A Northern Ireland study

In an unpublished study with 141 13–14-year-olds in Northern Ireland (reported in Greer, 1997), three paper-and-pencil tests were prepared, prominently labeled "Mathematics Test", "Estimation Test" and "Mathematical Puzzles", the idea being that the last two labels might, respectively, legitimize responses other than the routine "single correct answer" and alert students to look beneath the surface. The Mathematics Test consisted of a mixture of P-items (including the runner and the flask item) from Greer's (1993) original study, and S-items. The Estimation Test included 3 of the same P-items plus 3 further items which clearly were estimation tasks (e.g., estimating how many copies of a small circle would fit inside a larger circle without overlapping). Finally, the test labelled "Mathematical Puzzles" included also 3 of the P-items used in the Mathematics Test plus three standard mathematical puzzles (e.g., counting the number of triangles in a complex figure).

Thus, the variation in the presentation of the tests might have been expected to increase the number of RRs for the Estimation Test and Mathematical Puzzles test relative to the Mathematics Test for two reasons. First, the label given to the tests strongly suggests that the nature of the expected responses is very different. Secondly, the nature of the other items (clearly estimation tasks and puzzles, respectively) provides a strong hint, in the first case, that an estimated, rather than exact, answer is legitimate, and in the second case, that it may be necessary to be careful in interpreting the task. For example, the flasks item was common to all three tests. The Estimation Test invites an estimated answer taking account of nonlinearity of the rise in water depth as the flask tapers, rather than one based on direct proportionality. The label "Mathematics Puzzles" offers at least a potential hint that the answer may not be as straightforward as might appear at first sight. In the event, the alternative settings of questions made very little difference. On average, compared to the Mathematics Test, the percentage of RR responses was 11% higher on the Estimation Test (not a statistically significant difference) and was even slightly lower on the Mathematics Puzzles test.

Afterwards, short interviews were carried out with 12 of the students, and two trends were noticeable. First, many of the students were able to articulate an awareness of a difference between conventional answers expected in the context of school mathematics, and answers appropriate to real situations (as was the case in the study by Caldwell, 1995, reported in Chapter 2, pp. 24–28). Second, the interviews suggested that relevant contextual information (such as that the water would rise more quickly as the flask narrowed, or that the runner tires with distance) was generally known to the students.

Effects of task demand characteristics: A Dutch study

Van Lieshout, Verdwaald, and Van Herk (1997) compared the answers of 202 pupils from two regular elementary schools and three special schools for mildly retarded children on a paper-and-pencil test consisting of ten S- and ten P-items. Half of the ten P-items were taken from the study by Verschaffel et al. (1994) (namely the rope item, the runner item, the school item, the friends item, and the buses item), while the other five P-items were generated by the authors (e.g., "Joris and Pim live in the same house. They bike home together in 8 minutes. How many minutes must Joris bike when he bikes home alone?" or "Rachid and Sandra do a running competition. They start together. It takes each child exactly 4 minutes to get to the finish. How long did the competition last?". Contrary to Verschaffel et al.'s (1994) original study, participants received an introductory oral class instruction in which two example problems (one being a problematic one) were discussed and during which the instructor explained and showed that several kinds of responses (even "I can't do this problem" responses) were absolutely legitimate. Using pupils' results on the S-items as a criterion, comparable groups of pupils from regular and special schools were formed, and afterwards the results on the P-items of these matched groups were analyzed and compared.

First of all, their findings confirm the results of the previous studies. While their results are a little bit less dramatic than those of Verschaffel et al. (1994) (probably because of a combination of the warning provided at the onset of the test and the removal of some of the most difficult P-items from Verschaffel et al.'s (1994) test), the number of NRs was still overwhelming. More than half of the P-problems elicited less than 20% RRs, and among these the school item and the friends item were the most difficult. Once again, pupils were rather successful on problems involving the interpretation of a division with a remainder (like the buses item). Finally, pupils from special schools solved significantly more P-items with an RR than those from regular schools (42% and 29%, respectively). The lack of a superiority of the pupils from the regular schools can be considered as an indication of the importance of the "didactical contract" (Brousseau, 1980, 1984, 1990, 1997; discussed in Chapter 5) with respect to doing word problems and, more specifically, as evidence that these pupils more easily and quickly learn to think and act according to the rules of this contract than less able ones. A related, but somewhat different, hypothetical explanation is that there are (subtle) differences between the rules of the didactical contract stipulating the place of real-world knowledge in word problem solving for children in regular schools as compared to their peers in special education.

Effects of various minimal interventions: A Swiss study

Like Yoshida et al.'s (1997) study, Reusser and Stebler's (1997a) investigation was not merely a replication of the study by Verschaffel et al. (1994) reported in Chapter 2. It also extended that study in several ways, all aimed at testing the hypothesis that simply alerting pupils in some way or another to the problematic nature of the P-items will lead to a significant increase in the number of RRs compared to the results reported by Greer (1993) and Verschaffel et al. (1994).

Extension 1
Six weeks after the first administration of the ten pairs of problems (see pp. 23–24) a reduced set of 17 problems (10 P- and 7 S-items) was again presented to the same three classes of 13-year-old Swiss students. This time each of the problems was accompanied by a set of questions which were printed on the back of each task sheet and which had to be answered while, or immediately after, solving each problem. The questions explicitly asked the students to evaluate their difficulties of understanding as well as to judge the solvability of each of the ten problems, and further they contained an explicit invitation to freely comment on the problems. Whereas the students of one class worked individually, those from two other classes worked in pairs on the problems and on the accompanying questions.

Reusser and Stebler (1997a) hypothesized that working on the problems in conjunction with the questionnaire would not only increase the likelihood of real-world knowledge becoming activated but would also make the students more sensitive to the problematic modeling assumptions of the P-problems used. Moreover, they anticipated an amplifying effect of having students working cooperatively rather than individually on these problems. Given the re-testing situation (all students solved the P-items for the second time), and therefore the expectation of a training effect, the overall increase of RRs between the two tests of 10% (from 22% to 32%) for the students working individually and of 12% (from 17% to 29%) for those working cooperatively was considered very small. Despite the accompanying questions and the pair work, between half and two-thirds of all responses of students still showed no sign of realistic mathematical modeling.

Extension 2

In the two classes from Extension 1 that were tested in pairs, a further small intervention involving problems P1 (friends item) and P5 (runner item) took place the day after the second testing. The goal of this intervention was "to sow the seeds of doubts about the students' solutions" (Reusser & Stebler, 1997a, p. 316). For each of these two problems, one of the experimenters asked a number of task-specific questions and provided some task-specific help. For the friends problem, the experimenter said: "Are you sure that your answer (almost all students had written $6 + 5 = 11$ or $6 + 5 + 2 = 13$) is really true? Look at the problem again closely and try to put yourself in Carl's and Georges' place. Suppose you and your friend invite guests for your birthday party. One of you invites 6, the other 5 people. Maybe each of you writes down the names of the friends you want to invite. Is there sufficient information in the problem text or do you need to know more to solve it. What additional information could help you to solve the problem?" Afterwards, the student pairs had five minutes to review their previous solution and decide whether they thought that their initial answer was right or wrong. A similar procedure was used for the runner item, where students were also first asked if they were sure that their solution ($10 \times 17 = 170$ sec) was correct and then were explicitly invited to put themselves in John's place and imagine that they run 100 meters in 17 seconds and asking them to think how long it would take them to run 1 kilometer. As a result of this intervention, a considerable number of pairs changed from unrealistic to realistic reactions. For the P1 problem the number of RRs changed from 1 pair to 9 pairs (out of a total of 21) and for the P5 problem, the number of RRs went from 3 to 11 (out of a total of 18). Nevertheless, as these numbers reveal, a substantial number of pairs stuck to their original NR even after this strong pointer in a (more) realistic direction.

After this small intervention a whole-class discussion took place. During this conversation the pupils were asked to state their opinion about the following:

- Why were the problems solved without anyone wondering whether they can be solved at all?
- Why did many pupils think about the difficulties but not mention them?

Among the responses given by the pupils were the following:

- "I did think about the difficulty, but then just calculated it the usual way. (Why?) Because I just had to find some sort of solution to the problem, and that was the only way it worked. I've got to have a solution, haven't I?"
- "I suspected that it wouldn't work but solved it anyway. (Why?) Because otherwise the task wouldn't have been solved."
- "We thought it was an arithmetic problem. There just has to be a solution."

- "Simply did not think of it."
- "I am not sure why I did not wonder about it. That's actually quite strange."
- "It would never have crossed my mind to ask whether this task can be solved at all."

So, while some reactions (for instance, the first two in the above list) suggest that some students had struggled with the cognitive conflict of responding in a (more) realistic or a (more) "scholastic" way, the reactions of others suggest that they did not experience such a cognitive conflict simply because it did not enter their head to question the solvability of the problem or the validity of their straightforward solution.

Extension 3

In their third and most systematic extension, Reusser and Stebler (1997a) used an almost identical set of problems as in Verschaffel et al. (1994) and their own replication study (see pp. 23–24). Participants were 439 secondary school students aged 13 from 41 randomly selected seventh grade classes from three different types of school. In order to selectively facilitate activation of real-world knowledge, the conditions under which the problems were solved by the students varied in the degree to which the problematic mathematical modeling assumptions were signaled. Each student received a booklet containing 16 problems – the 10 P-problems from Verschaffel et al. (1994) and 6 S-items – in one of the following experimental conditions:

- Condition 1, in which a procedure identical to that used by Verschaffel et al. (1994) was followed.
- Condition 1a, in which four P-items (the items about the runner, the school distance, the rope and the flask) were slightly reformulated by adding a contextual sentence that should alert the students to a possible difficulty with the problems (e.g., the sentence "Make a sketch before solving the next problem" for the school problem, and the sentence "Study the picture carefully" for the flask-item).
- Condition 2, in which each problem was accompanied by a set of questions explicitly asking for the evaluation of its quality (e.g., difficulty of understanding, solvability). Contrary to the alerting additions from condition 1b, these questions were the same for all problems.
- Condition 3, in which, as well as the questions from condition 2, students were explicitly told to be cautious by means of the following bold printed text placed at the beginning of the test: "Be cautious. Some of the following problems are not as easy as they seem. There are, in fact, some problems in the booklet for which it is very questionable if they are solvable at all".

The main results were as follows. First, the average level of NRs remained remarkably high. Even though the number of RRs in this study with seventh

Table 3.1   Percentages of Realistic Reactions (RRs) to the Runner, the School, the Rope, and the Flask Items* and Mean Numbers of Realistic Reactions (RRs) in Conditions 1 and 1A (Reusser & Stebler, 1997a).

| | % of RRs to four P-items | | | | Mean number of RRs |
	Runner	School	Rope	Flask	
Condition 1	39	31	22	24	1.28
Condition 1A	49	35	33	48	1.70

*   For the full text of the P-items listed, see Table 2.2. For a detailed description of the criteria for scoring a reaction as NR or RR, see Verschaffel et al. (1994).

graders had more than doubled in comparison to the Verschaffel et al. (1994) study and to their own replication study (see pp. 23–24), still more than 50% of the observed reactions to the set of P-problems were unambiguously non-realistic. Second, a comparison between the three major signaling conditions (namely conditions 1, 2, and 3) revealed similar percentages of RRs: 42%, 39% and 41%, respectively. An Analysis of Variance carried out on these results showed that these differences were not statistically significant. Third, a statistically significant increase of RRs due to an instructional factor was observed with respect to condition 1a, i.e., the four problems that had been slightly elaborated by adding a specific alerting sentence for each particular problem. An Analysis of Variance on the results for the students from condition 1 and 1a on the four problems for which a reformulated version was used in condition 1a, showed a significant effect of change in the experimental setting, indicating that students solved these four problems realistically more often when a problem-specific instructional signal indicated a possible complication. Percentages of RRs for these four P-problems (namely the runner, the school, the rope and the flask items) are given in Table 3.1.

In sum, while even very explicit, general warnings to the students about equivocal, indeterminate, or even unsolvable problems did not increase their realistic behavior, alerts associated with specific single problems led to a significant but still moderate increase in the number of RRs.

Effects of two different scaffolds: A Flemish study

Finally, we report a follow-up study by Verschaffel et al. (1999) assessing whether two rather simple forms of scaffolding during an individual interview would be sufficient to transform children's non-realistic responses into realistic ones. This study consisted of two stages. In the first stage, seven of the ten P-items used in Verschaffel et al.'s (1994) first study together with their corresponding S-items were collectively administered to a group of 64 fifth-graders from three different schools (see pp. 23–24). The seven P-items selected were:

P1 (the birthday item), P2 (the planks item), P4 (the buses item), P5 (the runner item), P6 (the school item), P9 (the rope item), and P10 (the flask item) (see Table 2.2). The administration of the test and the scoring of the answers were done in the same way as in Verschaffel et al.'s (1994) first study, and the results of this collective test are described in Chapter 2 (pp. 18–23). Based on the results on this paper-and-pencil test, the five most "realistic" and the five most "non-realistic" problem solvers from each of the three fifth-grade classes were selected to participate in the second stage of the investigation. During this second stage,

Table 3.2   First Scaffold (S1) and Second Scaffold (S2) Used for the Seven P-items in the Study of Verschaffel, De Corte, and Lasure (1999).*

P1 (Friends item)
S1: One of your classmates said that it is impossible to solve this problem. Who is right? You or your classmate?
S2: Can you give me the name of a good friend in the class? Imagine you and ... are giving a party together and that you both invite your best friends. Imagine that your five best friends are present, and that the six best friends of ... are present. Are you sure that there are 11 guests at the party?

P2 (Planks item)
S1: One of your classmates responded in this way: "4 × 2 = 8; Steve can saw 8 planks". Who is right?
S2: Can you draw the planks? Can you also draw what happened with these planks according to the story. Can you see on this drawing how many planks of 1 meter Steve can saw out of these 4 planks?

P4 (Buses item)
S1: One of your classmates responded in this way: "450 ÷ 36 = 12.5, so they will need 13 buses". Who is right?
S2: You have answered that they will need 12.5 buses. What does that answer mean to you – 12.5 buses?

P5 (School item)
S1: One of your classmates said that it is impossible to solve this problem. Who is right? You or your classmate?
S2: Can you make a drawing of the situation described in the problem? Is this the only drawing you can make out of this problem statement?

P6 (Runner item)
S1: One of your classmates responded that it is impossible to know precisely how long it would take John to run 1 kilometer. Who is right?
S2: Do you know your best time on the 100 meters? Imagine that it is 17 seconds. If you had to run 1 kilometer, do you think that you would succeed in running every remaining 100 meters in that same best time?

P9 (Rope item)
S1: One of your classmates responded in the following way: "I don't know exactly, but the answer will certainly be more than 8 pieces". Who is right?
S2: Here is a picture of the man who is tieing together the pieces of rope to stretch between the two poles. Do you still think that 8 pieces will be enough to stretch between the two poles?

Table 3.2 continues

Table 3.2 (continued)

P10 (Flask item)
S1: One of your classmates said that it is impossible to give a precise answer to that question. Who is right?
S2: If this is the part of the flask that was filled after 10 sec (interviewer points at the shaded part of the cone-shaped flask on the pupil's response sheet), can you indicate on this figure what the level of the water will be after 20 sec? And after 30 sec?

---

*   For the full text of the P-items, see Table 2.2. For a detailed description of the criteria for scoring a reaction as NR or RR, see Verschaffel et al. (1994).

---

which took place one or two days later, these 30 problem solvers were individually administered the same seven problem pairs once again.

To assess the strength of children's tendency towards non-realistic responses, the following interviewing procedure was followed with respect to each P-item solved with a non-realistic answer (NR) during the collective paper-and-pencil test. First, the pupil was asked to read aloud the problem followed by his own NR written down on the answer sheet (e.g., "$4 \times 2.5 = 10$ planks" for the planks item). Then a cognitive conflict was provoked by confronting the pupil with the written notes of a fictitious classmate who had responded to the same problem in a realistic manner. For instance, with respect to the planks item the interviewer said: "As you can see on this sheet, one of your classmates responded: $4 \times 2 = 8$ planks. What is the best answer? Why?". If the pupil stuck to the initial NR (i.e., "10 planks") after this first scaffold, a second scaffold in the direction of realistic modeling was provided. The interviewer said: "Can you draw the planks? Can you also draw what happened with these planks according to the problem statement? Can you see on your drawing how many planks of 1 meter Steve can saw out of these 4 planks?". In Table 3.2 the first and the second scaffolds for each of the seven P-items are presented separately.

In order to control for possible confounding effects of the interviewing technique, the same procedure (involving similar kinds of provocation and scaffolding) was applied with respect to the P-items that were solved with an RR (e.g., answering the "planks" problem with "$4 \times 2 = 8$ planks") during the paper-and-pencil test. In these cases, the pupil was confronted with the fictitious written reaction of a non-realistic responder (e.g., "$4 \times 2.5 = 10$ planks"). The same interviewing procedure was also applied with respect to the seven unproblematic S-items. Pupils who had given a correct answer to an S-item from the paper-and-pencil test were confronted with an incorrect response resulting from the application of a wrong operation with the given numbers (e.g., a multiplication instead of a division), while those who had given an incorrect answer were confronted with the correct one.

For all 30 pupils and for all 7 P-items the following kinds of data were used for analysis:

- The initial reaction to the problem during the paper-and-pencil test (scored as RR or NR).
- The reaction to the confrontation with the RR of a fictitious classmate (weak scaffold) in the case of an initial NR during the paper-and-pencil test (also scored as RR or NR).
- Eventually, the reaction to the second and stronger form of scaffolding (again scored as RR or NR).

A first hypothesis of the study was that (as in the study of Verschaffel et al. (1994)) the overall number of RRs generated on the 7 P-items of the paper-and-pencil test would be very low. More specifically, it was predicted that this over-all percentage would not differ substantially from the percentage of RRs found in that initial study (i.e., 17%). Second, a positive effect of the scaffolds on the number of RRs was anticipated. Therefore, it was predicted that the percentage of RRs of the 30 pupils at the end of the individual interview would be signifi-cantly higher than their percentage on the paper-and-pencil test. However, it was expected at the same time that this overall percentage of RRs at the end of the individual interviews would still be dramatically low. This latter prediction was based on the hypothesis that pupils' tendency towards routine-based and non-realistic responding would be so strong and resistant that the confrontation with the two scaffolds would frequently still be insufficient to make them change their initial NR into a RR.

In line with the first hypothesis, the results on the collective test did indeed reveal a very strong tendency among the pupils to exclude real-world knowledge and context-based considerations. Only 16% of all the reactions of the 64 pupils to the 7 P-items of the paper-and-pencil test were classified as realistic. This per-centage was almost exactly the same as that in Verschaffel et al.'s (1994) first study (17%). (For details about the percentages of RRs for each P-item, see Table 2.3.) As said before, from each of the three classes in which the paper-and-pencil test was administered the five pupils with the highest number of RRs and

Table 3.3   Absolute and Cumulative Frequencies and Percentages of Realistic Reactions of the 30 Pupils to the Seven P-items at the Different Stages of the Individual Interview (Verschaffel, De Corte, & Lasure, 1999).

		IRR*	RRF	RRS	NRR
Absolute	N	49	34	37	90
	%	23	16	18	43
Cumulative	N	49	83	120	210
	%	23	39	57	100

*   IRR = immediate realistic reaction during the collective test
     RRF = realistic reaction after the first scaffold
     RRS = realistic reaction after the second scaffold
     NRR = still non-realistic reaction after both scaffolds

the five pupils with the lowest number of RRs were selected to participate in the second part of the study.

In line with the second hypothesis, a statistically significant effect of the two forms of scaffolding was found. Altogether, the two scaffolds resulted in an increase in the cumulative number of RRs from 23% (i.e., 49 RRs out of a total of 210 responses) during the paper-and-pencil test to 57% (i.e., 120 RRs out of a total of 210 responses) at the end of the individual interviews (see Table 3.3). A one-tailed t-test revealed that this increase was statistically significant (p < .001). Additional one-tailed t-tests revealed that both forms of scaffolding contributed equally to this increase. The first scaffold resulted in an increase of the cumulative percentage of RRs from 23% (i.e., 49 RRs out of 210 responses) to 40% (i.e., 83 RRs out of 210 responses) (p < .001), and the second and stronger scaffold produced an additional significant increase from 40% (i.e., 83 RRs out of 210 responses) to 57% RRs (i.e., 120 RRs out of 210 responses) (p < .001).

Although the scaffolds led to a significant increase in the number of RRs, the percentage of cases in which the first and second scaffolds resulted in a shift from a NR to a RR was smaller than the percentage of cases in which the scaffolds did not have any effect on pupils' NR (see Table 3.3).

Summary and discussion

In this chapter, a number of follow-up studies have been reviewed that tested the effectiveness of minimal interventions intended to make students more alert, to sensitize them to the consideration of aspects of reality, or to legitimize alternative forms of answer without fundamentally changing the experimental setting. The results of these investigations suggest that these minor variations in the experimental setting may produce, at best, weak effects. This conclusion holds especially for scaffolds not specifically tied to particular problems. When item-specific scaffolds were applied, as in Verschaffel, De Corte, and Lasure's (1999) study and in some of the extensions of Reusser and Stebler (1997a), the results were somewhat better, but were still small. In any case, these studies lead to the conclusion that students' manifest lack of attention to realistic considerations when doing word problems in school, as documented in the earlier studies reported, cannot be explained completely, or even mainly, in terms of (experimentally induced) lack of alertness on the part of the students or of their refusal to respond with forms of answers other than a single, precise, numerical answer because they expected that such answers were not allowed or would not be welcomed by the investigators. In the next chapter, studies about the impact of another, more drastic, kind of alteration of the experimental setting are discussed.

4

Increasing the Authenticity of the Experimental Setting

RA (= research assistant): Board of Education Central Receiving Office, may I help you?
T(homas): Yes, I'd like to order 7 minivans.
RA: O.K., may I have your name please?
T: Thomas.
RA: What school are you calling from?
T: Louis Armstrong.
RA: What date do you want the minivans?
T: I'd like them on Friday, April 15.
RA: Can you tell me where the children will be going?
T: Ricardo's Restaurant, Queens.
RA: How many children will be transported?
T: 32
RA: Why do you need 7 minivans?
T: Um, I arrived to have 7 minivans by... dividing 5 into... 30 by 5 and I got 6 and there was a remainder 2 so I need 7 because the buses won't allow, it's against the law to have more than 5 kids.
RA: O.K., would you like to order anything else?
T: No, thanks.
RA: O.K., thank you, good-bye.
T: Bye.
(DeFranco & Curcio, 1997, p. 66)

The investigations reviewed in Chapter 3 revealed the relative ineffectiveness of minimal interventions intended to make students more alert, to sensitize them to the consideration of aspects of reality, or to legitimize alternative forms of answer. In this chapter we will review another set of follow-up studies in which the role of changing a more fundamental aspect of the problem situation was investigated, namely its authenticity. We are aware that the term "authenticity" has no single and simple meaning. If applied to mathematical application problems, the term is related to contrasts in various aspects of the ways in which word problems are administered to and experienced by students in the context of a mathematics lesson, on the one hand, and the ways in which people deal with mathematically equivalent (isomorphic) problems situated in (real-life) settings outside school, on the other hand (Cooper, 1992, 1994; De Lange, 1995; Lajoie, 1995; Lesh & Lamon, 1992a; Lave, 1988, 1992). These various aspects relate to:

- The way the problem is encountered by the solver, e.g., imposed by someone else versus posed by the solver him/herself.
- The nature of the solver's goals, e.g., success or at least survival in the classroom versus solving a truly engaging dilemma.
- The formulation of the problem, e.g., pre-formulated and accompanied with all the requisite data versus loosely defined and with the information necessary for solving it to be actively sought from a variety of sources.
- The social and material conditions under which the problem has to be modeled and solved, e.g., without versus with help of others and/or of particular cultural tools.
- The criteria by which the solution will be judged – merely by criteria of a mathematical nature or also by non-mathematical criteria some of which may be political, moral or social.
- The consequences of the modeling activity, e.g., a good or bad mark from the teacher versus an immediate and unequivocal reaction from the social and/or material environment on the solver's intervention based on the result of his modeling activity.

The more the task setting resembles the second alternative on each of these aspects, the greater the authenticity of the task.

In the studies reviewed in this chapter various aspects of enhancing the authenticity of the task are discussed. While, in some studies, the changes relate to making the task presentation more meaningful and appealing to students, others focus on increasing the amount of discussion and negotiation that is allowed, and still others on the pedagogical context within which the mathematical task is encountered.

A more authentic version of the buses problem

A first example of a study in which researchers have experimented with the effectiveness of the presentation of mathematical wor(l)d problems in a (more) authentic setting, was carried out by DeFranco and Curcio (1997). The starting point of this study was students' difficulty with division problems with a remainder (DWR), such as the buses item, due to their failure to map the computational result back to the real-world context of the problem, as documented in the study of Silver et al. (1993) (see pp. 6–8), and also in the studies of Greer (1993), Verschaffel et al. (1994), and others (see Chapter 2). Starting from such observations, DeFranco and Curcio examined students' interpretation of remainders in division as applied to similar word problems embedded in two different experimental settings – one being a restrictive scholastic setting, and another one being a (relatively) real-world setting. In the first part of the study, 20 sixth-graders were confronted with the following version of the buses item in a restrictive context (i.e., an individual interview in which pupils were questioned about mathematical word problem solving): "328 senior citizens are going on a trip. A bus can seat 40 people. How many buses are needed so that all the senior citizens can go on the trip?". In the second part, the same 20 pupils were asked to make a telephone call using a teletrainer obtained from the telephone company to order minivans to take sixth-graders to a class party. The (oral) request to make a phone call was accompanied with a fact sheet with relevant information, as shown below. Since this task placed children in a (more) reality-based situation, the context in which it was presented was considered (more) "real-world" by the experimenters.

Facts:
Date of party: Friday, April 15
Time: 4:00–6:00 PM
Place: Ricardo's Restaurant, Queens
Number of children attending the party: 32

Problem:
We need to transport the 32 children to the restaurant so we need transportation. We have to order minivans. Board of Education minivans seat 5 children. These minivans have 5 seats with seatbelts and are prohibited by law to seat more than 5 children. How many minivans do we need? Once you have determined how many minivans we need, call 998–2323 to place the order.

Only two of the 20 children responded to the buses item appropriately in the restrictive setting. Of the 18 children who produced an inappropriate response, 17 made an incorrect interpretation of the remainder (e.g., by responding with an answer involving a remainder or by rounding their result down to 8 buses without any further comment). In the real-world setting, 16 out of the 20 students gave an appropriate response. Thirteen of them ordered 7 minivans because they realized part of a vehicle could not be ordered, and the other 3 ordered 6 minivans but gave reasonable reasons for doing so (e.g., they explained that another, smaller vehicle "like a car or something" would be needed to transport the remaining two students). Thus, although this study focused on only one particular kind of P-item from Greer's (1993) and Verschaffel et al.'s (1994) problem set, its results nevertheless strongly support the idea that by embedding this particular kind of P-problem in a (more) authentic setting, many students will be led away from solving it in a stereotyped and non-realistic way.

Reasoning realistically in meaningful communicative projects

The effectiveness of increasing the authenticity of the context was illustratively investigated for one particular kind of P-problem by Wyndhamn and Säljö (1997). While in the previous study the focus was on students' difficulties in interpreting their computational answer for DWR problems, Wyndhamn and Säljö (1997) worked with another item from Verschaffel et al.'s (1994) test, namely the one about school distances (P6).

Starting from a sociocultural analysis of the critical features of the experimental setting used in the investigations of Greer (1993) and Verschaffel et al. (1994), these authors argued that the origins of the difficulties in making realistic considerations and giving realistic answers documented in these studies should be sought in students' assumptions (or beliefs) regarding how to read and interpret messages. Therefore, attempts to elicit more realistic reactions from students should focus on introducing elements in the experimental setting (or, in Greer's (1997) terminology, changes in the experimental contract) that contribute to making it more natural or reasonable for them to pay attention to the referential meaning of statements from word problems and to meaning-based ways of thinking while doing these problems. So, the main hypothesis was that if students were to become involved in meaningful communicative projects with others, the likelihood of realistic considerations and reactions would increase. To test this hypothesis, an empirical study with a problem similar to one of the P-items from the test of Verschaffel et al. (1994), namely the school-distance problem, was carried out. Besides that problem, another problem about the distance between two towns was used:

- Anna and Berra attend the same school. Anna lives 500 meters from the school and Berra 300 meters from the school. How far apart from each other do they live?
- What is the distance between Alstad and Broby according to these road signs?

The two distance problems were not given as explicitly formulated word problems in either written or oral format. Instead, they were introduced within a conversation in which three students and one researcher took part. The conversation in which the first task (the school-distance problem) was introduced concerned the roads to school for the participants in the discourse. Together with the problem phrasing, a fictitious map of a school with its surroundings was shown, but there was no indication on the map of possible locations of the homes of Anna and Berra. The second problem (about the road signs) was introduced as part of a general discussion about road signs – their colors, where one would find them, how they are to be interpreted, and so on. This problem was accompanied by a sheet with a figure showing two road signs indicating that the distance to one city (Broby) was 17 miles to the right, and the distance to another city (Alstad) was 8 miles to the left.

Problem 1 was given to eight groups of 10–11-year-olds of equal ability, and problem 2 to six other groups of 12-year-olds. Each group discussed its problem for about 15 minutes. The most clear-cut outcome of this study was that all groups, except one, arrived at the conclusion that it was impossible to give one clear-cut answer to these problems and that "it all depends". This outcome contrasts sharply to what was found in the original study of Verschaffel et al. (1994) and its several replications, in which very few or no RRs involving such context-based considerations were observed. The only exception to this pattern was one group of high performers who finally insisted on giving one single answer for the school-distance problem, in spite of the fact that they had realized during the discussion that there could be more than one correct answer.

In sum, the study of Wyndhamn and Säljö (1997) shows that making realistic considerations and using these considerations when modeling and solving mathematical application problems is certainly within the "zone of proximal development" (Vygotsky, 1986) of most students. Apparently, it suffices to disentangle the problems from their scholastic chains and to embed them in another, more meaningful and more authentic setting to transform apparently non-realistic and stereotyped problem solvers into clearly realistic and thoughtful ones.

The same investigators have carried out several other studies yielding additional confirmatory evidence in favor of this conclusion (Säljö & Wyndhamn, 1993, 1988a, 1988b). Earlier, in 1993, these authors had conducted an experiment with 214 eighth- and ninth-grade students of a Swedish comprehensive school (15–16- years olds) who were asked to determine the cost of a domestic letter of 120 grams in two different settings – in the context of either a mathematics lesson or of a lesson called "social sciences" (Säljö & Wyndhamn, 1993). In both settings, subjects received a copy of the official postage table of the Swedish Post (as shown in Figure 1.1). The participants solved the problem individually as a paper-and-pencil task, and the analyses were based on the written solutions and explanations presented by the students. The students were not asked to perform the authentic action of sending a letter.

The major goal of the study was to analyze which type of interpretation and solution procedure would be applied by the students – attempting to read the postage from the table, or using it as a tool for calculating (e.g., by adding up the postages for a weight of 20 grams and a weight of 100 grams or by multiplying the postage for a weight of 20 grams by 6). Based on the hypothesis that the context of the mathematics lesson would discourage students from paying attention to the referential meaning of numerical information, the authors expected that the "reading off" procedure would be used more frequently and the calculating approach would be used less frequently in the social sciences setting as compared to the mathematics lessons setting.

The results supported this hypothesis. When given the task of establishing the postage rates for letters in the context of a mathematics class, 57% of the students interpreted this as a mathematics task, thus engaging in some kind of calculations in order to come up with an answer. When dealing with the same problem in the social studies class, only 29% used mathematical operations to arrive at an answer. Moreover, there was an obvious difference between the two contexts in terms of the groups of participants who read the table to find the postage rate. The proportion of students who chose the lower rate (4.00 kroner) tended to be higher among those who dealt with this task in the mathematics class than among those who got it as a social-sciences task. According to the authors, this was due to the fact that in the mathematics class students typically have to round off to the closest quantity, which, in this case, is the lower value. Thus, even here, the mistake made may, when considered within the context of the discourse situation, turn out to be the consequence of the specific communicative conditions under which the students attempted to find the postage rate.

To summarize, the above-mentioned studies of Säljö and Wyndhamn indicate that the overall communicative context in which students find themselves tends to determine their interpretation and solution of the task. Dealing with problems about distances or postage rates as part of a mathematics lesson has a tendency to lead to students perceiving the task as an invitation to perform calculations on the numerical data given in the problem statement without taking into account their available experiential knowledge about these problem situations, while, conversely, when the task is embedded outside the mathematical context (e.g., in the context of a social or environmental studies class), there is a tendency to abstain from (immediate) calculating and to use context-based knowledge to represent and solve these problems.

Realistic mathematical modeling in the context of performance tasks

Starting from anecdotal evidence of striking contrasts between individual students' non-realistic modeling and problem-solving behavior when confronted with P-items in the typical context of school arithmetic (see Chapters 2 and 3), on the one hand, and the appropriate and successful problem-solving endeavors of the same students on certain "performance-based" tasks in the Swiss part of the Third International Mathematics and Science Study (TIMMS), Reusser and Stebler (1997b) set up a study in which they analyzed students' solutions of P-problems in two different contexts, namely:

- The same, restricted, school-like context as in the studies of Greer (1993), Verschaffel et al. (1994) and their own replication study (see pp. 23–24).
- A contrasting context in which the same P-problems were presented in the form of performance-based tasks.

The hypothesis was that students would show much less evidence of non-realistic mathematical modeling and inadequate problem solving, and thus produce many fewer non-realistic answers (NRs), in the second context as compared to the first one. To test this hypothesis, Reusser and Stebler (1997b) transformed five out of the ten P-items from the paper-and-pencil test of Verschaffel et al. (1994) and their own replication study (Reusser & Stebler, 1997a) – i.e., the planks item, the runner item, the school item, the rope item and the flask item – into performance tasks. This was done by presenting the (still verbally formulated) problems together with concrete materials (planks, a saw and a meter-stick for the planks problem; pieces of rope, scissors and a meter-stick for the rope-problem; a real cone-shaped flask and a water jug filled with liquid for the flask-item, etc.), and with a clear performance instruction. This instruction consisted of the following elements:

- Investigate the problem using the concrete materials accompanying it.
- Make a prediction about the answer.
- Execute the task and comment on what you are doing.
- Decide whether you will stick to your initial prediction or if you want to change it.
- Write down your final answer.

The first day each student solved the five P-items in the traditional paper-and-pencil format, and the next day they solved the same tasks individually in the performance setting.

In line with their hypothesis, Reusser and Stebler (1997b) found a remarkable and significant increase of RRs from the paper-and-pencil condition to the performance setting condition on four of the five P-problems. For the planks item, the runner item, the rope item, and the flask item, the percentage of RRs changed, respectively, from 25% to 56%, from 12% to 47%, from 18% to 62%, and from 7% to 40%. Only for the item about the school distances did the increase in the percentage of RRs from the restricted (21%) to the performance-based setting (36%) fail to reach statistical significance. A closer analysis of students' problem-solving activities revealed that both the mere confrontation with the concrete materials, and the effective manipulation of these materials fulfilled a scaffolding function toward (more) realistic mathematical modeling.

Given students' significantly better performances in the more authentic settings, the authors take their results as positive evidence for their hypothesis that it is hardly the inability of the students, but rather the type of task and experimental context that must be seen as the primary cause for students' unrealistic answers to P-items as documented in the studies of Greer (1993), Verschaffel et al. (1994) and their own previous work (Reusser & Stebler, 1997a). Although we endorse Reusser and Stebler's (1997a) interpretation of the results of their study, it should still be recognized that still about half of the responses of the P-items were non-realistic notwithstanding the researchers' efforts to enhance the authenticity of the problems.

Summary and discussion

The studies reviewed in Chapter 3 showed that minimal variations in the experimental setting intended to sensitize students to the consideration of aspects of
reality by means of a general warning, or to legitimize alternative forms of
answers, may produce, at best, weak effects. The effects of the changes in the
experimental context reported in the present chapter, by contrast, reveal much
greater improvements in students' achievements on the P-items, and, more specifically, in their inclination and their capability to include the real-world knowledge
and the realistic considerations they were so reluctant to activate under the previous, more restricted testing conditions. In other words, these findings suggest
that when the nature of the "premises for the interactive ritual" (Wyndhamn &
Säljö, 1997, p. 379) – or, as Greer (1997, p. 305) has called it, the "experimental contract" – afford it, students are prepared and able to take realistic considerations into account when responding to mathematical problems.

In this respect, it should be stressed that in none of the experiments reported
in this chapter were children put in problem situations that were presented in a
completely authentic out-of-school setting, as was done in the work of ethnomathematicians such as Nunes et al. (1993) or Saxe (1988). In DeFranco and
Curcio's (1997) study, the (more) authentic setting was a simulation of a real
phone call (on a teletrainer) and not making a real telephone call to order transportation for a school trip. In their investigation about the school-distance problem, Wyndhamn and Säljö (1997) confronted pupils with a *drawing* of a road
sign (rather than putting them on a crossing with road sign) and the accompanying question about the distance between the two cities on the road sign was
raised in a setting that was still a scholastic one. Similarly, in Säljö and
Wyndhamn's (1993) research about the postage rates problem, children were
asked to figure out how much postage to put on a letter of a certain weight in
two different school settings (i.e., a mathematics versus a social sciences class)
– they were not supposed to perform the actual action of posting a letter. Finally,
Reusser and Stebler (1997b) compared students' performance on a paper-and-
pencil test with that on a performance-based test, not within a leisure activity
wherein they were actually buying and sawing planks for a bookshelf, or
knotting together some pieces of rope to get a longer one, or filling differently
shaped flasks with beverages for a birthday party. Rather than being involved
in truly authentic problem situations (or "quantitative dilemmas" as Lave
(1992) calls them) participants in these studies were still attending to representations and simulations of everyday actions and problem situations, and their
major motivation for attending to these problems was one of learning or performing as defined in a certain formal setting (and not ordering buses, or sending letters, or "bricolaging" with planks, ropes, and flasks). So, while all these
studies examined alternative assessment settings or formats that might target
more directly the need for sense-making and solution interpretation (see also De
Corte et al., 1996), they still did not involve a comparison of students' attention
to, and use of, real-world knowledge when interpreting and solving application

problems in a school mathematics setting versus in a genuine real-world setting. Reflecting upon this aspect of their postage rate study, Säljö and Wyndhamn (1993, p. 335–336), commented as follows:

> Had the problem appeared in a different, real-world setting as part of a genuine decision of what it would cost to send a letter to a friend, other strategies of seeking the relevant information would have been available, including the most natural one of asking someone who knows ... or to put on extra stamps because it is better to be on the safe side rather than running the risk of embarrassing the person receiving the letter, who would have to pay extra and maybe even collect the letter at the post office... In the world of formal schooling, such modes of reasoning that are sensitive to the social meaning of the act of sending letters would probably be regarded as attempts at evading the problem given.

Besides these studies of DeFranco and Curcio (1997), Wyndhamn and Säljö (1997), and Reusser and Stebler (1997b), there is also an older line of research about making word problems more authentic that we have not discussed heretofore in this chapter, namely earlier work on the effects of "concretizing" or "personalizing" word problems. These studies range in focus from the broad examination of concrete versus abstract examples (see, e.g., Caldwell & Goldin, 1979) to the personalization of problem contexts (see, e.g., Anand & Ross, 1987; Davis-Dorsey, Ross, & Morrison, 1991; Ross, 1983).

Personalization can be realized by adapting the context of the problem applications to students' backgrounds and interests; for instance, Ross (1983) compared performance on statistical problems that were embedded in either an educational or a nursing context by students majoring in education versus nursing. Alternatively, the computer can be used to replace fictional problem characters and problem situations with descriptions of actual people and events based on the student's biographical information; this was, for example, done in a study with elementary-level arithmetic materials and pupils by Anand and Ross (1987).

Taken as a whole, this research provides consistent evidence that learners are more successful at solving word problems that convey themes or contexts oriented around familiar events, people, or activities. According to Ross and associates, this success is due, first of all, to the fact that the personalization generates greater interest in the task, and thus increased attention and effort.

A second possible function of the personalization is making the problem themes or contexts more meaningful in the sense of being easier to relate to existing knowledge schemata. Although this research demonstrates that students perform better when the word problems are personally meaningful and attractive to them, the nature of the problems used in these studies, the kind of experimental design used, and the performance-based nature of the data analysis, limit their relevance for the present analysis. Indeed, the results of these studies do not yield direct evidence that changing the experimental context results in a drastic improvement in students' perception of and approach to word problems, and, more specifically, in their inclination to include the real-world knowledge and the realistic considerations they are so reluctant to activate under more restricted, traditional testing conditions.

Part 2

The Educational Environment: Analysis and Reform

In Chapter 2 we presented a set of closely related studies by Greer (1993), Verschaffel et al. (1994), and others, on upper elementary and lower secondary school children's conceptions about and use of real-world knowledge in doing school arithmetic word problems. The findings from these studies indicated that, when confronted with mathematical applications in a typical school setting, the vast majority of students demonstrated a strong tendency to solve arithmetic word problems in a stereotyped way by applying one (or a combination) of the basic arithmetic operations with the given numbers in the problem, without paying attention to possible problematic modeling assumptions underlying their proposed solution from a realistic point of view. As such, these findings extended earlier observations made by other investigators about students "suspension of sense-making" reported in Chapter 1.

A first series of subsequent studies, reported in Chapter 3, showed that small variations in the experimental setting aimed at sensitizing students to the consideration of aspects of reality and/or legitimizing alternative forms of answer at the onset of the test, are not enough to transform students' unrealistic answers into more realistic and meaningful ones. In Chapter 4, another set of studies was reviewed showing that, in contrast to the relative ineffectiveness of simply alerting students to the problematic nature of problems, a more radical approach, namely framing the problems within more authentic contexts, has a dramatic positive effect on the realistic nature of students' responses.

While in Part 1 we concentrated mainly on reporting observations, we were, of course, considering deeper theoretical issues from the start. For example, among the lines of investigation for further research suggested by Greer (1993, p. 248) at the end of his pioneering study about this topic, were the following:

> How important is the social/educational context in which the items are presented?
> Are students aware of the implicit rules of the word-problem "game"?
> How easy would it be to change the beliefs and conceptions underlying students' responses by appropriate teaching?
> How aware are teachers of the issues raised?

Further, in the discussion section of the report on their first study, Verschaffel et al. (1994, p. 292) wrote as follows:

> ... attention has to be paid to the variety and complexity of the target problems to which pupils are exposed in teaching and testing situations ... More specifically, the impoverished diet of standard word problems offered in traditional mathematics classrooms should be replaced – or at least supplemented – by a wide variety of more authentic and more complex problem situations that stimulate or require pupils to pay attention to realistic modeling ... However, simply exposing pupils to a wide variety of authentic and complex problem situations may not be enough. These problems should be embedded in a powerful instructional environment that explicitly aims at the development of the proper concepts, skills and attitudes that are needed for realistic modeling of problem situations and for realistic interpreting of outcomes of arithmetic calculations, as part of a genuine mathematical disposition ...

In this second part of the book the issues identified in these two quotations are addressed.

Chapter 5 considers key factors in the instructional environment that we believe underlie the observed patterns of student behavior, beginning with a detailed analysis of the nature of word problems as typically taught in schools and the (mainly implicit) rules of the "word problem game", and the factors that influence students' initiation into, and participation in, that game. A number of major contributory factors may be identified, including the way in which word problems are dealt with in textbooks and in assessment instruments, both of which are strong sources of signals helping to shape the rules of the word problem game for both students and teacher. As adumbrated by Greer (1993, p. 248), a major direction for future research is to switch attention from students to teachers, whose role is, of course, of paramount importance. The chapter concludes with a first step in this direction, namely a detailed study of preservice teachers' conceptions and beliefs about word problems (Verschaffel, De Corte, & Borghart, 1997).

In Chapter 6, we move from mainly ascertaining studies – designed to document the existing state of affairs in terms of students' and teachers' behavior when answering word problems – to intervention studies designed to see to what extent a radical alteration in the instructional environment can change student responses and, by implication, their underlying conceptions. The results of two teaching experiments carried out in one of our research centers (Verschaffel & De Corte, 1997b; Verschaffel, De Corte, Lasure, Van Vaerenbergh, Bogaerts, & Ratinckx, 1999) are reported. These studies provide promising illustrations of instruction designed to promote more sophisticated understanding of word problems as exercises in modeling aspects of the real world. They are supplemented by a description of another exemplary instructional environment that significantly addresses issues that we have identified, namely the work of the Cognition and Technology Group at Vanderbilt (1997). This chapter ends with a description of similar efforts to anchor the assessment of mathematical problem solving in more realistic contexts, and an example of how reform of assessment positively impacted teaching (Clarke & Stephens, 1996).

5

The Effects of Traditional Mathematics Education

Mother: Sir, I would like to make a remark about one of the mathe-
matical problems from my son's last homework. It's about the prob-
lem in which he had to set up an algebraic expression to find out the
price of one loaf of bread starting from a set of givens. As he could
not solve it himself, I helped him. In our first trial we arrived at a
price of 21.5 francs for a loaf of bread. I told my son that this could
not be the correct answer, and so we restarted our solution process,
and we arrived at a response which seemed more reasonable to us,
but which finally turned out to be wrong.

Teacher: But your initial answer of 21.5 francs was correct. I took
this problem from an older mathematics book. Of course, we all
know that nowadays a loaf of bread costs considerably more than
21.5 francs. But after all, that's not what students have to worry
about when doing algebra problems. It's the construction and exe-
cution of the mathematical expression that counts, all the rest is
decor.

(Excerpt from a discussion between a mother of a 13-year-old stu-
dent and her son's mathematics teacher at a parents' evening (Van
der Spiegel, personal communication, 1997))

As explained at the end of Chapter 4, the findings and observations reported in
that chapter clearly and convincingly demonstrate that sense-making and solution

interpretation are within the reach of most, if not all, students. In our view, what causes the students to react non-realistically in the restricted, scholastic context of the mathematics class is not their inability to take realistic considerations into account and to apply the criterion that what they do should make sense, but their interpretation of the "rules of the game" of the interactive ritual in which they are involved (De Corte & Verschaffel, 1985; Gravemeijer, 1997; Greer, 1997; Hatano, 1997; Reusser & Stebler, 1997b; Wyndhamn & Säljö, 1997). As soon as these rules or premises were relaxed (i.e., by removing the P-items from the restricted and routine context of a typical school arithmetic lesson or test and by placing them in a more authentic, stimulating and communicative setting) many students seemed to get rid of their inclination to approach these P-items in a stereotyped, meaningless way without paying attention at the realistic constraints which call into question the appropriateness of their routine solutions.

In the first section of this chapter, we will analyze more closely these rules and expectations that seem to lie at the basis of students' inclination to do school arithmetic word problems in a non-realistic way. Then we will reveal some aspects of current instructional and assessment practice and culture, which seem to be responsible for the origin and development of these student perceptions. The chapter concludes with a detailed study of pre-service teachers' conceptions and beliefs about word problems.

Students playing the game of school arithmetic word problems

The rules of the game of word problems

The belief of students that word problems are divorced from reality cannot be considered or investigated in isolation. According to several authors, it is part of a more general knowledge and belief system about word problems, the way these problems are structured and formulated, and the role they play in the teaching and learning of arithmetic at school. De Corte and Verschaffel (1985, p. 7) introduced the term "word problem schema" to refer to this system. These authors define it as a general mental schema, complementary to the more specific schemas about the semantic relations underlying particular kinds of word problems (e.g., the Change, Combine, and Compare schemas for representing one-step addition and subtraction problems). The word problem schema involves the solver's knowledge and beliefs about:

- The intent and role of word problems, intuitive understanding of which enables the solver to react appropriately to a word problem (as opposed to a joke, a riddle, etc.) even when the problem statement or the accompanying instructions contain very few explicit cues to that end.
- The typical structure of word problems, which orients the solver's reading process from the onset toward certain elements and relations that are crucial for the construction of an appropriate problem representation and

which leads him/her to disregard other pieces of information because of their irrelevance for that problem representation.

- A number of implicit rules, presuppositions, and agreements inherent in the game of word problems which enable the solver to resolve further ambiguities and obscurities in the problem text, and which further help him or her to react to the word problem as intended by the problem poser. Examples of such rules of the game are given below.

In a similar way, Lave (1992, p. 77) writes that:

> There is a discourse of word problems – a set of things everyone knows how to say about word problems or that can be expected in "word-problemese", issues and questions that come up when people begin to talk about them; and things that are not and cannot be said within this framework.

Several authors have carried out analyses of the underlying rules and expectations that need to be used by students (but also by all other participants in the game of word problems, such as textbook writers, test developers, teachers and parents) to make the game of word problems function properly, and that are assumed to be at the disposal of students having sufficient experience with this particular game (De Corte & Verschaffel, 1985; Gerofsky, 1996, Kilpatrick, 1985; Lave, 1992; Nesher, 1980; Reusser & Stebler, 1997a; Schoenfeld, 1991). Among the assumptions that are listed in these analyses are the following:

- Assume that every problem presented by the teacher or in a textbook is solvable and makes sense. Do not ask questions. Accept the correctness, completeness and meaningfulness of word problems from authority on trust.
- Assume that there is only one correct answer to every word problem, and that this has to be a precise and a numerical one.
- Assume that this single, precise and numerical answer can and must be obtained by performing one or more mathematical operations or formulas with the numbers in the problem, and almost certainly with all of them.
- Assume that the task can be achieved using the mathematics one has access to as a student – in fact, in most cases, by applying the mathematical concepts, formulas, algorithms, etc. recently encountered in mathematics lessons.
- Assume that the final solution, and even the intermediate results, involve "clean" numbers (i.e., whole numbers).
- Assume that the word problem itself contains all the information needed to find the correct mathematical interpretation and solution of the problem, and that no information extraneous to the problem may be sought. The problem should not be altered by importing contextually relevant information elements that are not explicitly given in the problem statement and that might complicate its intended underlying mathematical model.

- Assume that persons, objects, places, plots, etc. are different in a school word problem than in a real-world situation, and don't worry (too much) if your knowledge or intuitions about the everyday world are violated in the situation described in the problem situation.

This set of assumptions has both positive and negative aspects. If there were not some (implicit) rules, the use of word problems in school arithmetic would be impossible (as in the case of any other type of language game; see Greer, 1997). On the other hand, these rules, of which the participants in the game may be largely unaware, encourage and sustain routine-based, superficial thinking by all players in this particular language game.

Considering word problems as a language game with its own rules of play, hidden expectations, intelligent tactics, etc. brings us to the notion of "didactical contract". This notion was introduced by Brousseau (1984, 1990, 1997) as a theoretical necessity imposed by his effort to understand dysfunctions in students' mathematical learning. It refers to a set of interrelated rules and mutual expectations that is negotiated in the mathematics classroom and that dominates the interaction between the teacher and the students during the mathematics lessons as well as the cognitive processes of an individual student when engaged in thinking and problem-solving activities in or related to these lessons. Typically, the underlying control exerted by the didactical contract operates without the participants (including the teacher) being aware of the fact that their communicative and cognitive processes are so strongly shaped by these rules and expectations. Thus, this didactical contract dictates, "explicitly to some extent, but mainly implicitly" (Brousseau, 1997, p. 31), how students and teachers are expected to behave in a mathematics class, how they must think and communicate with each other, what sort of tasks the teacher can give to his or her students, what questions students are allowed to ask the teacher and what kind of responses the teacher is expected to give (and vice versa), and so on. In a paper published in 1980, Brousseau (1997, p. 225) defines the didactical contract as follows:

> In a teaching situation, prepared and delivered by a teacher, the student generally has the task of solving the (mathematical) problem she is given, but access to this task is made through interpretation of the questions asked, the information provided and the constraints that have been imposed, which are all constants in the teacher's method of instruction. These (specific) habits of the teacher are expected by the student and the behavior of the student is expected by the teacher; this is the didactical contract.

Cobb and his colleagues introduced the closely related concept of "sociomathematical norms" (Cobb, Yackel, & Wood, 1992; Gravemeijer, 1994, 1997; Yackel & Cobb, 1996). Other terms like "hidden curriculum" or "hidden agenda" are also sometimes used in this context (Puchalska & Semadeni, 1987).

One of the first phenomena to be interpreted in terms of the didactical contract was the set of findings relating to children's responses to the famous captain's age problem, and similar problems, discussed in Chapter 1. Already in 1983, Chevallard explained children adding the numbers of sheep and goats together in order to find the age of the captain as an effect of the didactical contract rather than as evidence of children's abnormality or stupidity (as the sensational press did), but Chevallard's text remained somewhat in the "gray literature" until 1988 when it appeared in a publication of IREM de Marseille (Chevallard, 1988; see also Brousseau, 1997; Joshua & Dupin, 1993). The following quotation is from Chevallard (1988, cited in Joshua & Dupin, 1993, p. 267):

> C'est le fonctionnement de ce dernier (= de contrat didactique) que les auteurs "mesurent" et non la "logique" supposée des enfants. Le contrat comporte en effet une clause valable pour tous les problèmes proposables dans le cadre didactique-scolaire, aux termes de laquelle: un problème proposable (légitimement) possède une réponse et une seule (acceptable au sens du contrat): pour parvenir à cette réponse toutes les données proposées doivent être utilisées, aucune autre indication n'est nécessaire et l'utilisation pertinente des données fournies se fait selon un schème mettant en jeu des procédures familières, au stade considéré, règles qu'il suffit alors de mobiliser et de combiner de manière adéquate, – ce qui constitue d'ailleurs le véritable champ de l'action de l'élève, sa marge de manoeuvre et d'incertitude. La manoeuvre realisée par les auteurs de l'enquête grenobloise constitue donc une rupture délibérée du contrat didactique usuel.

To support his claim, Chevallard (1988) pointed to the fact that a number of children in the Grenoble studies who answered the captain's and/or the shepherd's problem by adding, subtracting, multiplying or dividing the two given numbers expressed serious concerns about the appropriateness of the problem and/or about the meaningfulness of their solution in reaction to the question "What do you think of the problem?" which followed each problem, as in the following example: "I cannot see any relationship between the sheep and the captain … I find this problem somewhat bizarre … I think it is a stupid problem because at first they speak about sheep and then about the captain." Why did (some of) the children who showed evidence of such concerns nevertheless respond with a numerical answer to these absurd problems, Chevallard (1988) asked himself. His answer was as follows:

> C'est que ce n'est tout simplement dans son rôle dans le contrat. Celui-ci définit les droits et les devoirs des élèves, les droits et les devoirs de l'enseignant. Le second doit s'assurer que le problème posé a une réponse et une seule. L'élève, lui, tenant pour légitimement acquis que le problème a une réponse et une seule doit fournir la réponse attendue. Le contrat n'inclut pas dans la tâche de l'élève que celui-ci ait à contrôler la légitimité contractuelle du contrat qui lui est proposé. (p. 268)

In this respect, the following excerpt from one of Chevallard's (1988) interviews – reproduced by several other authors (e.g., Baruk, 1985; Puchalska & Semanedi, 1987) – is very illustrative:

> I: You have 10 red pencils in your left pocket and 10 blue pencils in your right pocket. How old are you?
> C: 20 years old.
> I: But you know very well that you are not 20 years old.
> C: Yes, but is your fault; you did not give me the right numbers!

Of particular significance here are also the conjectures made by Freudenthal (1982, p. 71; see also Puchalska & Semanedi, 1987; Selter, 1994) of a possible magical context some children are searching for when trying to solve the captain's problem:

> To solve a problem, they look for secret marks, for signals hidden by accident or intention, and, in particular, for numbers to put them on the right track . . . How old is the captain? The 26 sheep and 10 goats on board are like the data used by the astrologer to foretell the future.

Thus, some students tried to see the problems from a different perspective which would allow them to somehow connect the given numbers to the unknown quantity. Selter (1994, p. 36) gives the following (very) creative constructions of such connections, which he observed in his interviews with pairs of elementary school children being asked to solve a number of absurd problem about the captain's and the shepherd's age: "The shepherd was given a sheep or a goat on each of his birthdays" or "He bought one animal for each year of his life; so he always knows how old he is". The more Selter (1994) looked at the videos of the interviews with the pairs of children, the more rational thinking he found and the more he felt obliged to modify the accusatory attitude he had taken initially toward the children:

> As long as we searched for irrationality, we could find it. However, when we looked for a rational background to the answers offered, a lot of irrational comments, and ways of acting turned out to be quite reasonable.

> . . . Children's ways of thinking are more often more reasonable, more organized and more intelligent than we suppose them to be. What we notice is contingent on our interpretation of the way they are acting and on the pedagogical contract between them and us. (Selter, 1994, p. 37)

Empirical evidence

It is our conviction that the construct of "the game of word problems", as an instantiation of the more general theoretical notions of "didactical contract" or "sociomathematical classroom norms", is useful to explain why students responded to the P-items with non-realistic rather than with realistic reactions (see Chapter 2), why they continued to produce NRs after minimal interventions aimed to alter their non-realistic response tendencies (Chapter 3), and why they did exchange their NRs for RRs to a considerable extent when the same P-items were administered in a radically altered experimental context wherein this didactical contract was no longer effective (Chapter 4).

However, we acknowledge that the set of beliefs and expectations that together constitute the "word-problem schema" (De Corte & Verschaffel, 1985) – i.e., the individual-psychological counterpart of the social-anthropological notions of didactical contract or sociomathematical classroom norms applied to word problems – is a hypothetical construct for which we have not yet come up with much direct empirical evidence. The direct empirical evidence provided has consisted of some incidentally produced or experimentally elicited explanations of students for their non-realistic responses to P-items, wherein they articulated (some of) the above-mentioned rules of the game of word problems.

Further evidence for the theory that school children gradually construct conceptions and beliefs about word problems as a genre is provided by contrasting the unexpected incorrect reactions (by the rules of the game) on school word problems made by subjects who are not (yet) familiarized with that genre, such as beginning elementary school children or unschooled adults, with the almost complete absence of such errors in subjects with considerable experience with the game of word problems. Illustrations of such unexpected reactions produced by young children can be found in earlier work of Verschaffel (1979, 1984; see also De Corte & Verschaffel, 1985) in which first-graders were collectively or individually administered a set of one-step addition and subtraction word problems in the beginning or the middle of the first grade (i.e., at a stage in the instructional process when they had barely encountered word problems yet).

- In a study of Verschaffel (1979), a first-grader was given the following word problem as part of a post-test: "In the evening Pete has 6 marbles. During the day he lost 2 marbles. In the morning Pete had...", and rather than filling in the numerical response "8 marbles" he completed the final sentence with the words "played with them".
- In Verschaffel's (1984) longitudinal study about the development in first-graders' skills in solving additive word problems, items like "Pete had some apples. He gave 3 apples to Ann. Now he still has 5 apples. How many apples did he have in the beginning?" were frequently answered by "some apples", "a couple" or "a few" during the first and the second interview (in the beginning and in the middle of the first grade).
- During the first interview of that same study, a bright first-grader showed great difficulty in solving the following problem: "Pete had 3 apples. Ann gave him 5 more apples. How many apples does Pete have now?", which was considered the easiest item from the whole problem set. The reason why she could not solve this problem was that after she had (materially)

represented the first sentence by taking three blocks, she decided that it was impossible to (materially) represent the second sentence *because the problem did not (explicitly) state in the beginning of that sentence that Ann had apples too.* From the fact that in the beginning of the story only the number of apples that Pete possesses was mentioned, this girl inferred that the other person in the story (Ann) had no apples. This is not an unreasonable conclusion for someone who is not familiarized with the game of word problems. Once the interviewer had explained that she could give Ann whatever number of apples she liked – for instance, a whole pile of apples – the girl was able to solve the problem.

In De Corte and Verschaffel's (1985) view, all these unexpected reactions resulted from children's lack of understanding of the intent, the structure and the style of word problems, and thus from their ignorance of what was expected from them in reaction to the presentation of such a strange piece of text.

For examples of inappropriately informal interpretations and solutions of word problems made by unschooled adults, we refer to Luria's (1976) famous work comparing the performance of illiterate and literate adults on a series of cognitive tasks including arithmetic word problems. These studies, conducted in the 1920s and early 1930s in a remote part of Soviet Central Asia, included illiterate adults who were still engaged in traditional modes of production and a second group of adults who had already been exposed to radical restructuring of the socio-economic system (e.g., collectivization) and culture (e.g., literacy). On all tasks, Luria (1976) found considerable differences between both groups. Whereas illiterates displayed a mode of thinking that relied heavily on personal experiences and practical considerations, literate subjects demonstrated a more abstract or theoretical attitude that allowed them to disregard particular daily-life experiences and shift into hypothetical reasoning unsupported by concrete conditions. For example, in one arithmetic word problem subjects were asked to calculate the difference between two towns (A and C) given information about the distance between each town and an intermediate town (AB and BC) that deviated from or contradicted subjects' practical knowledge about the actual distances between these three towns. Whereas literate subjects proceeded without concern about the accuracy of the information provided, non-literates refused to reason within the (contrafactual) conditions given in the problem and preferred to rely on arguments and reasonings based on their everyday concrete experiences. The following is an excerpt from an interview with Khamrak, a thirty-six-years-old peasant from a remote village, almost completely illiterate (Luria, 1976, p. 129):

> I: From Shakhumardan to Vuadil it is three hours on foot, while to Fergana it is six hours. How much time does it take to go on foot from Vuadil to Fergana?
> K: No, it's six hours from Vuadil to Fergana. You are wrong. It's far and you wouldn't get there in three hours.
> I: That makes no difference. A teacher gave this problem as an exercise. If you were a student, how would you solve it?

> K: But how do you travel? On foot or on horseback?
> I: It's all the same. Well, let's say on foot.
> K: No, then you won't get there. It's a long way ... If you were to leave now, you'd get to Vadil very, very late in the evening.
> I: All right, but try and solve the problem. Even if it's wrong, try to figure it out.
> K: No. How can I solve a problem if it isn't so.

Remember that all the unexpected (from our perspective) reactions mentioned above were typically from subjects with little or no experience in school mathematics in general and with the socio-cultural device of word problems in particular. Therefore, the disappearance of such reactions among students with some years of enculturation in the genre suggests that, after some time, most of them seem to have a rather well-developed schema with regard to word problems which includes knowledge of many facets of the genre, including the knowledge that it is not expected from them to try to make sense of the story in terms of everyday life, to search for deficiencies or contradictions in the story, and so on.

Further evidence for the fact that most children seem to develop rather quickly most facets of the word-problem schema is provided by contrasting inexperienced beginning elementary school childrens' inability to construct word problems themselves, as documented by De Corte and Verschaffel (1985; see also Brissiaud, 1988), with older pupils' success in generating such problems. Menon (1993) documented that upper elementary school pupils who were asked to formulate their own mathematical questions produced very nearly exclusively word problems in canonical forms. Besides his own data, Menon (1993) also refers to Ellerton's (1989) large-scale study of 10,000 secondary students in Australia and New Zealand who, when asked to write one difficult word problem, overwhelmingly wrote problems similar in form to those in their textbooks. Lave writes in this respect:

> If you ask children to make up problems about everyday math, they will not make up problems about their experienced lives; they will invent examples of the genre: they too know what a word problem is. (Lave, 1992, p. 77)

Like Lave, we interpret the discrepancy between young and inexperienced pupils' inability to generate well-formed word problems (in terms of the usual standards of the genre) and older and more experienced pupils' capacity to produce conventionally correct word problems, as another direct piece of evidence in favor of the assertion that after some years of involvement in the game of word problems children have absorbed most of the assumptions underlying that game.

This assertion, particularly in respect of the rule that persons or objects behave differently in word problems and in real-life situations, is nicely illustrated in the never-ending wave of jokes about word problems circulating among school children. Typical examples are the following: "Six birds are sitting in a tree; suddenly two of them are shot by a hunter; how many birds are still sitting on that tree? Answer: none, because all the other birds flew away immediately after hearing the gunshot", and "Mother has to divide 10 apples equally among her 8 children. How could this be done? Answer: By making applesauce from these 10 apples", and many others. The fact that these jokes are so popular among school children suggests that they, at least intuitively, grasp the divorce between word problems and problem solving in real life.

To summarize: using examples from the research literature on word-problem solving, we documented how children slowly but surely build up a schema with regard to word problems as they get more and more experience with the genre, which governs their acting and thinking in instructional and/or assessment contexts wherein this schema is considered to hold. The important point we tried to make here is that the findings reported in all previous chapters of this book about schoolchildren answering word problems, with apparent scarce regard for whether their answers make sense when considered from a realistic viewpoint, must also be interpreted from this perspective. Indeed, the explanation of these findings must be sought initially through analysis of word problems as a peculiar genre, more generally by a consideration of the cultural context of the classroom, and, most generally of all, through a consideration of the nature of schooling itself, as is elaborated in Part 3 of this book. Stated differently, students' responses to word problems that apparently disregard considerations of reality should be interpreted as showing that they are adhering to conventions learned and reinforced over a considerable period of time. From this socio-cultural perspective, it seems inappropriate to describe students' behavior with respect to P-problems as "irrational", or as "blind and deaf concerning given data" or even as "mentally abnormal", and certainly unjustified to blame the students for their irrational behavior and their chloroformed minds, as was done in some initial analyses of the captain's problem and in the sensational press (see Chapter 1). Rather, the students can be considered to have been behaving rationally (Selter, 1994) "in accordance with their socialization in school practices" (Wyndhamn & Säljö, 1997, p. 379) or, as Brousseau (1984, 1990, 1997) would put it, as stipulated in the didactical contract.

This brings us to several key questions. By what learning processes are these rules of the game of word problems (and especially the rules about the unrealistic nature of these problems) internalized by children? Why are they so resistant to change? Are they explicitly and deliberately installed and maintained by the curricular materials and/or the teacher? In answer to the last question, we would say typically not, although we are aware of some painful illustrations of explicit interventions by the textbook materials and/or teachers aimed at establishing rules for word problems and their intentions and assumptions. Take, for instance, a Flemish textbook series wherein the key words in word problems are emphasized by printing them in a specified color (De Corte, Verschaffel, Janssens, & Joillet, 1985) or the example of the teacher presented at the start of this chapter who explained to the mother of one of his students that word problems may have unrealistic numbers and, accordingly, that students should not ask themselves whether their obtained numerical answer makes sense from a realistic point of view. As with most other aspects of the didactical contract (Brousseau, 1984, 1990, 1997) or of socio-mathematical classroom norms (Cobb et al., 1992), the development of students' views of word problem solving as an activity with artificial rules and without any specific relation to out-of-school reality occurs implicitly, gradually and tacitly through being immersed in the culture of the mathematics classroom. More specifically, this enculturation seems to be mainly due to the following two aspects of the current instructional practice and culture in which children learn to solve mathematical application problems.

- The impoverished and stereotyped diet of standard word problems occurring in mathematics lessons and tests.
- The way in which these problems are conceived and treated by teachers.

Stereotyped and artificial nature of word problems

Students' beliefs about the intent, the structure, the style and the underlying assumptions of school mathematics word problems, and the superficial and artificial problem-solving tactics accompanying these beliefs, are first of all enhanced by the nature of the problems with which pupils are confronted day-by-day in the mathematics lessons and tests. Overviewing the literature on this topic, the current problem base is considered as having the following characteristics.

First, especially in the early grades of the elementary school, pupils often receive problems that can be solved by the application of one of the four basic arithmetic operations to the numbers mentioned in the problem. This, of course, creates and reinforces the belief among pupils that every application problem can be solved in this way. In an analysis by De Corte et al. (1985) of a representative sample of six frequently-used mathematics textbooks for the first grade of the Flemish elementary school, it was found that the vast majority of the story problems belonged to this category. Stigler, Fuson, Ham, and Kim (1986) compared mathematics textbooks for grades 1 and 2 in the U.S. and the Soviet Union, and found that the vast majority of the problems in the U.S. textbooks were one-step problems, in contrast to the textbooks used in the Soviet Union in which two-step problems occurred much more frequently (up to 37% in grade 1 and 53% in grade 2). Reusser (1988) observed that, while in the middle and upper grades of the Swiss elementary school the number of multi-step problems increases, complex problems requiring more than 2–3 solution steps and solution times of more than a couple of minutes remain extremely rare. It seems reasonable to infer that these characteristics of the problem set strengthen, among other things, students' beliefs that solving a word problem consists of a small number of solution steps which never takes more than 5 minutes.

Second, in traditional mathematics textbooks word problems are presented, grouped and formulated in a way that makes the application of superficial, routine-based strategies (like looking at cues in the chapter headings or in the environment in which the problem appears or searching for key words in the problem which will tell you what operation to perform) undeservedly successful (Sowder, 1988). Säljö and Wyndhamn (1987) noted that Swedish mathematics textbooks often contain headings that clearly spell out the nature of the tasks to be performed (e.g., "Multiplication"), so that pupils know what operation to perform even before they have started to read the problem itself. In their comparison of mathematics textbooks for grades 1 and 2 in the U.S. and the Soviet Union, Stigler et al. (1986) found that U.S. textbooks frequently had whole pages of a single kind of word problem, in contrast to textbooks used in the Soviet Union in which word problems on a page were quite varied in structure. An analysis of German textbooks by Stern (1992) revealed that most of the non-trivial word problems occurring in these books required a subtraction with the given numbers. This led her to the conclusion that a pupil who follows the rule "Whenever there is a problem I don't understand immediately, I subtract the numbers" will succeed on 90% of the problems presented in these textbooks. Schoenfeld (1991) found that in some U.S. textbooks the vast majority of the word problems (in some textbooks up to 90%) could be solved successfully by means of the key-word strategy. This strategy consists of executing the operation that is associated with the key word in the problem statement (e.g., "When the problem contains the word 'altogether' then add; when it contains the word 'left' then subtract, etc."), – a finding that was also obtained in De Corte et al.'s (1985) analysis of Flemish textbooks for grade 1. Schoenfeld (1991) provided also some evidence of the detrimental effect of this key-word strategy on children's solutions of word problems by referring to elementary school pupils who erroneously subtracted the two given numbers because of the presence of a person called "Mr. Left" in the problem statement.

Third, traditional textbooks contain few or no problems including superfluous, misleading or missing data. Thus, all numbers needed to solve the problem are explicitly stated in the problem and every given number has to be used. From these characteristics of the problem set, students quickly, easily, and reasonably infer that solving word problems is "doing something with (all) the given numbers". In their analysis of mathematics textbooks for grade 1, De Corte et al. (1985) found no problems with superfluous or missing data, apart from a couple of problems featuring a price list of particular wares (e.g., sweets or school supplies) and a list of questions of the form "What would it cost Pete to buy X and Y?". Similarly, Stigler et al. (1986) found that in U.S. textbooks for grades 1 and 2 word problems with extraneous or missing data were very rarely included. Puchalska and Semadeni (1987), on the other hand, report that the official curriculum in Poland encourages teachers of grade 1–3 to give occasionally problems with missing, surplus or contradictory data, and that these problems are scattered in some Polish textbooks (and see Krutetskii, 1976).

Fourth, although the numerical tasks are embedded in a context, the stereotyped nature of these contexts, the lack of lively and interesting information about the contexts, and the nature of the questions asked at the end of the word problems jointly contribute to children not being motivated and stimulated to pay attention to, and reflect upon, (the specific aspects of) that context. Rather, these characteristics of the problem formulation and presentation signal the students to consider this context as an incidental and irrelevant "dressing" of the mathematical task that is presented through the medium of the context, and to solve this mathematical task by neglecting contextual complications and nuances. For instance, based on an analysis of typical items from recent national assessment instruments in the U.K. and their accompanying scoring systems (including a problem about how many times a lift that can carry only 14 people would have to go up to take 269 people – a problem resembling the buses item), Cooper (1992, 1994) concluded that these test items are typically intended to be treated as closed, in the sense that reference should not be made outside the given information to what might be considered relevant everyday knowledge. Accordingly:

> ... it appears to be the case that success in school, as measured by national testing in mathematics, will depend on the child's capacity or willingness to approach tasks with a particular orientation to meaning, one which brackets out everyday, common-sense knowledge as a resource. (Cooper, 1994, p. 162)

What makes Cooper's (1992, 1994) analysis so interesting, is that it points to the incompatibility between, on the one hand, the embedding of these test items in narrative contexts (as a result of the legitimization of "practical", "problem-solving" and "investigational" approaches by, in particular, the Cockcroft Report, 1982), and the requirement for the child to avoid any move into common-sense reasoning, on the other (Cooper, 1994).

Fifth, students are regularly confronted with word problems that (presumably unintentionally) contain more or less extremely unrealistic conditions, numbers and questions. Typical examples of problems in which certain real-life constraints are seriously violated are those about cars driving or animals running for hours at a regular speed, gardens that are completely fenced off (with no way in or out), people reading book pages at a regular pages/hour or pages/day speed, and so on. Other problems contain given numbers (of distances, prices, speed ...) which are ridiculously far removed from their counterparts in the real world (as illustrated by the example mentioned in the beginning of this chapter), or unrealistically "clean". In still other problems, the question raised at the end of the story is one that would not naturally arise for someone actually in that particular situation. From these characteristics, students learn that there is a gap between school word problems and the real world and that, in order to succeed on word problems, it is better not to refer to the real, and possibly even to bracket out their knowledge of it.

Sixth, word problems requiring estimation rather than exact computation are rare in traditional elementary school textbooks and tests (Cooper, 1992, 1994;

Sowder, 1992; Treffers & De Moor, 1990). This absence of estimation tasks cultivates the belief that the correct answer to a word problem is necessarily a single, precise number, and that other types of answers (e.g., conditional answers, answers expressed as a range of possible answers, approximate answers, etc.) are less valuable or even completely illegitimate. Yet many tasks that are unrealistic (e.g., cars travelling at constant speed for many hours) become reasonable if presented as estimation problems.

Seventh, indeterminate, equivocal, or unsolvable problems are very rare in traditional mathematics textbooks and tests (Greer, 1997; Reusser, 1988). From the fact that these textbooks don't confront children with problems that allow multiple interpretations, alternative situational models, various solution paths and different answers, pupils construct the belief that there is a one-to-one relation between every word problem and its single correct interpretation, solution, and answer.

Eighth, situations requiring the formulation of problems (starting from some given elements), the generation of structurally similar or different problems (from a given target problem), and the sorting of problems (rather than solving them, etc.) occur rarely in current instructional practice (Brown & Walter, 1993; English, 1998; Kilpatrick, 1987; Silver, 1994). Students invariably have to solve application problems given to them by others. This is very different from the real world, in which identifying, defining and formulating the problem is a genuine part of the problem-solving process.

In an attempt to summarize (some of) the characteristics of traditional word problems which lie at the basis of students' beliefs about and strategies for solving word problems discussed above, Reusser and Stebler (1997a, p. 323) wrote as follows:

> Only a few problems that are employed in classrooms and textbooks invite or challenge students to activate and use their everyday knowledge and experience. Most word problems used in mathematics instruction are phrased as semantically impoverished, verbal vignettes. Students not only know from their school mathematical experience that all problems are undoubtedly solvable, but also that everything numerical included in a problem is relevant to its solution, and everything that is relevant is included in the problem text. Following this authoring script, many problem statements degenerate to badly disguised equations.

Earlier in this chapter (pp 63–64) we provided some empirical evidence showing that, in contrast to school beginners, students with some years of (traditional) mathematical experience have at their disposal a set of beliefs, expectations and attitudes about arithmetic word problem solving, whereby it is viewed as an activity with artificial rules and as divorced from out-of-school reality. In the first part of this section, we have reviewed different characteristics of the problem base that elementary school children confront in traditional instruction and assessment of word problem solving, some of which character-istics have been demonstrated by research. However, strong and systematic empirical evidence *directly* supporting the central claim – that the nature of the word problems offered to students in instructional and assessment settings is directly responsible for the observed lack of sense-making among school child-ren – is lacking. We are aware of only one exception.

As part of their study reported in Chapter 3 (pp. 37–38), Reusser and Stebler (1997a) included at the end of each test booklet a set of eight questions where-by students were asked about their classroom experience with four kinds of mathematical tasks: (1) unsolvable tasks, (2) undetermined tasks, (3) tasks with more than one sensible solution, (4) tasks with a solution that can only be esti-mated. While an Analysis of Variance showed no significant effect on the num-ber of RRs to P-items of (subjectively reported) classroom experience with respect to tasks with more than one solution, or to estimation tasks, highly statistically significant relations were found between the observed level of RRs and the self-reported classroom experiences with unsolvable and undetermined tasks. Students who more strongly indicated (on a four-point Likert scale) having already dealt in class with tasks either being unsolvable or containing important gaps of information produced significantly more RRs than their colleagues indi-cating less or no such experience. Although this small study can be criticized for relying on students' self-reports rather than on a systematic study of the materi-als that these students actually encountered during their school career, it provides at least some empirical evidence for the effect of the nature of the problems set on students' beliefs about and approaches to word problem solving.

Effects of assessment

Another major component of the instructional environment that has an influ-ence on children's and teachers' uptake of the word problem game is the nature of standard forms of assessment.

At the most general level, it has been pointed out that assessment influences mathematics instruction because it transmits powerful signals conveying the goals of instruction, what counts as competence in mathematics, and what forms of mathematical performance are valued. Indeed, according to Clarke (1996), assessment is one of the major means by which the didactical contract is communicated. He asserts, moreover, that "students infer the goals of the cur-riculum and the criteria for quality with greater confidence from assessment than from the rhetoric of the teacher" (p. 346).

The invidious position in which even the best-intentioned of teachers find themselves is well expressed by the following (Bell, Burkhardt, & Swan, 1992a, p. 119):

> The implementation of higher-order thinking in the school mathe-matics curriculum depends on the provision of appropriate

assessment material. Teachers' natural and laudable desire to see students succeed at public examinations is bound to be reflected in their teaching. Short, closed, stereotyped examination questions are bound to encourage imitative rehearsal and practice on similar tasks in the classroom (WYTIWYG or "What You Test Is What You Get").

In terms of generating, and responding to, mathematical tasks – word problems, in particular – there are many features of standard pencil-and-paper assessments that militate against the setting of more authentic tasks and the possibility of other than routine responses. These features include:

- *Working under time constraints.* Standard written assessments are typically timed, and often in such a way as to produce pressure (which has the convenient effect, from a narrow psychometric point of view, of increasing the range of scores). Students who stop to think about possible complications or alternative models are penalized because this more thoughtful approach takes time.

- *Denial of access to other information.* The text must be assumed to contain all the information relevant to formulating a solution.

- *Lack of opportunity to experiment.* In the course of written tests, students cannot investigate with manipulatives, other physical objects, or software and other modeling tools.

- *No interaction with other students.* There is no chance to discuss the relative merits of alternative models, for example.

- *No communication between the assessor and assessed.* If a task posed appears ambiguous or unclear to the student, (s)he generally has no opportunity to seek clarification; conversely, the assessor has no opportunity to probe the students' answers. Consequently, there is a premium on routine answers – a student that introduces complications, such as realistic considerations, may be penalized.

In short, the typical written assessment is closed – in terms of time, in terms of information, in terms of activity, in terms of social interaction, in terms of communication. These features encourage the use of questions that are fragmented and stereotyped and help to establish and reinforce the assumptions about word problems listed on page 59. Indeed, the same limitations as listed above apply, to a considerable extent, to many students' experience of working on word problems in class (though it doesn't have to be like that).

The evidence for the detrimental effects of assessment on teaching and learning mostly is circumstantial, based on observation of the typical state of affairs and logical inferences therefrom. It is hard to see how the most direct form of experimental evidence, involving comparison groups of students and teachers working under contrasting assessment regimes, could be forthcoming, particularly since for a proper test, the systems would need to be in place for several years. The closest feasible approach to an experimental test is to monitor the changes in teaching following changes in assessment, and an example of such a study is considered in the next chapter, in the context of efforts to reform assessment.

Teacher beliefs and teacher behavior with respect to word problem solving

In conjunction with the nature of the word problems encountered by children, the way in which these problems are conceived and actually treated by teachers in mathematics lessons is another plausible explanatory factor for the development in students of beliefs about and tactics for word problem solving that lie at the basis of the results reported in previous chapters. Indeed, it seems reasonable to assume that the teachers' personal views on the major aims of doing word problems, and, more particularly, their own disposition toward realistic mathematical modeling, will influence several aspects of their teaching behavior such as: (1) the problems they select from textbooks or generate themselves for students, (2) the kind of comments they make and the kind of instructions they give with respect to the problems they administer to students, (3) how they react to any critical comments or questions about (the unrealistic nature of) the problems, (4) the amount of discussion about (problematic) modeling assumptions they allow and even encourage in their mathematics class, (5) the kind of feedback they provide on students' answers and thinking processes, especially to responses with manifest traces of (un-)realistic mathematical modeling, (6) the kind of problems they select or generate for informative and summative mathematics tests (see previous section). It can be assumed that all these aspects of the teachers' behavior will in turn influence students' learning, and especially the beliefs, expectations, attitudes and tactics they develop with respect to word problem solving. Whereas there is research evidence about the relationship between teachers' mathematical beliefs, their behavior in the mathematics classroom, and their students' mathematical learning processes and outcomes in general (see De Corte et al., 1996; Fennema & Loef, 1992; Thompson, 1992), we do not know of ascertaining studies in which the teachers' actual (pedagogical) content knowledge or their actual teaching behavior with respect to realistic mathematical modeling has been systematically investigated and related to student learning processes and outcomes.

There is one study that sheds some light on this issue, namely an investigation by Verschaffel et al. (1997), which analyzed (future) elementary school teachers' conceptions and beliefs about the role of real-world knowledge concerning the problem context in the modeling of school arithmetic word problems, as reflected in:

- The pre-service teachers' own spontaneous responses to a set of word problems with problematic modeling assumptions.
- Their evaluations of (imaginary) pupil answers to these problems that do or do not take into account relevant real-world knowledge.

Although subjects in this study were future rather than actual teachers, and although the study does not involve a direct linkage between these student-teachers' beliefs and their actual teaching behavior and/or student outcomes, we nevertheless report it here because it yields at least some information about (future) teachers' cognitions and beliefs with respect to teaching and learning word problem solving.

Design

Participants were 332 pre-service elementary school teachers from three differ-ent teacher training institutes in Flanders. About two-thirds were pre-service teachers who had just started their first year of training, while one third were third-year students who had almost completed their pre-service training. A paper-and-pencil test was constructed consisting of 14 word problems: seven non-problematic standard items (S-items) and seven problematic items (P-items). The seven problematic items (P-items) selected were as used in previous studies: P1 (the friends item), P2 (the planks item), P4 (the buses item), P5 (the runner item), P6 (the school item), P9 (the rope item), and P10 (the flask item) (see Table 2.2).

The test was given twice to all pre-service teachers, but each time with a dif-ferent task. The first time (Test 1), the student-teachers had to answer the 14 word problems themselves. Calculations and comments could be written down in a comments box below the answer box (see Figure 5.1). Immediately after they had finished and handed in this test, they were given Test 2, in which they were asked to score four different answers from pupils to the same 14 word problems as in Test 1 with either 1 point (absolutely correct answer), 0 points (completely incorrect answer) or ½ point (partly correct and partly incorrect answer). The four response alternatives to the seven P-items in Test 2 belonged to four different categories: a non-realistic answer (NA), a realistic answer (RA), a technical error (TE) and another answer (OA) derived by using the wrong operation or giving one of the numbers in the problem (see Figure 5.2). At the bottom of each problem, there was a box for writing explanations and/or comments.

For Test 1, an analysis of the student-teachers' spontaneous solutions to the P-items was carried out involving two major categories: realistic reaction (RR) versus non-realistic reaction (NR). The analysis of the pre-service teachers' reac-tions to the seven P-items in Test 2 focused on the score (1, ½ or 0) given to the realistic answer (RA) and the non-realistic answer (NA).

Hypotheses and questions

Hypotheses and questions with respect to Test 1
The first hypothesis with respect to Test 1 was that – due to their continuing experience with an impoverished diet of standard word problems and to the lack of systematic attention to the mathematical modeling perspective in their mathematics lessons – even the student-teachers would demonstrate a strong

The flask is being filled from a tap at a constant rate. If the depth of the water is 3.5 cm after 10 seconds, about how deep will it be after 30 seconds?

Correct answer:

Comments:

Fig. 5.1   Presentation of the problems in Test 1 (Verschaffel et al., 1997, p. 342).

The flask is being filled from a tap at a constant rate. If the depth of the water is 3.5 cm after 10 seconds, about how deep will it be after 30 seconds?

Answer A: $3 \times 3.5 = 11.5$ cm. After 30 seconds the depth of water will be 11.5 cm.

Score: ....

Answer B: $3 \times 3.5 = 10.5$ cm. After 30 seconds the depth of water will be 10.5 cm.

Score: ....

Answer C: 3.5 cm + 20 cm = 23.5 cm. After 30 seconds the depth of water will be 23.5 cm.

Score: ....

Answer D: It is impossible to give a precise answer

Score: ....

Comments:

Fig. 5.2   Presentation of the problems in Test 2 (Verschaffel et al., 1997, p. 342).

tendency to exclude real-world knowledge when confronted with the problematic versions of the problems, and, consequently would solve these problems as if they are not at all problematic. Although it was expected that the results of these 18–21-year-old pre-service teachers would be considerably better than those of the upper elementary school and lower secondary school children from the studies of Greer (1993), Verschaffel et al. (1994), and others, a low percentage of RRs on the seven P-items from Test 1 was predicted.

Second, the experimenters wondered whether there would be a difference between the students who had almost completed their (third year of) training as elementary school teachers and the students who had just started their (first year of) training. On the one hand, one could assume that the third-year students would have a stronger disposition toward realistic mathematical modeling since, during their theoretical and practical courses in mathematics education, these third-year student-teachers must inevitably have run into situations that invited or necessitated dealing with and reflecting upon the difficulties involved in realistic mathematical modeling. On the other hand, one could also suppose that the quantity and/or the nature of these occasions may be insufficient to have a significant positive impact on these student-teachers' cognitions and beliefs about the issue of realistic mathematical modeling. One even could argue that due to their three years of experience with a traditional approach toward elementary school mathematics in general and with school arithmetic word problems in particular, these third-year student-teachers might have developed a stronger disposition toward stereotyped, non-realistic modeling and interpreting of arithmetic word problems than the first-year students. Because of the existence of these conflicting hypotheses, no specific prediction was made about the difference in the percentage of RRs between the first-year and the third-year students.

Hypotheses and questions with respect to Test 2
A first hypothesis was that the student-teachers' scores of the realistic (RA) and the non-realistic answer (NA) to a P-item in Test 2 would generally reflect their spontaneous solutions of the same problems during Test 1. Consequently, it was anticipated that the student-teachers would frequently consider the NA for the seven P-items in Test 2 as the (perfectly) correct answer and accordingly score it as 1, whereas the RA would frequently be conceived as an inappropriate response and accordingly scored as 0. Second, the investigators wondered whether there would be a difference in the scores for the NAs and the RAs between the first-year and the third-year students. However, for the same reasons explained above, no specific prediction was made about the strength or the direction of this difference.

Hypotheses and questions with respect to the relationship between Test 1 and Test 2
A final set of hypotheses and questions concerns the relationship between a student-teacher's answer to a P-item in Test 1, on the one hand, and his or her score for the NA and the RA on the same item in Test 2, on the other hand. Generally speaking, a fairly good match was predicted between the student-teachers' reactions to each of the seven P-items on the two tests. This implies the following two predictions. First, most subjects who answered a P-item from Test 1 with a NR,

will give a score of 1 for the NA and a score of 0 for the RA for the same item in Test 2. Second, an RR on a P-item during Test 1 will typically be followed by a score of 0 for the NA and 1 for the RA for the same item in Test 2. However, the experimenters also anticipated a number of possible learning and thinking processes in the student-teachers that could lead to a mismatch between a student-teacher's own response to a P-item during Test 1 and the way in which (s)he scored the realistic and the non-realistic response alternatives for the same item in Test 2. A first possible source of such mismatches is the scaffolding effect of the confrontation with a realistic answer. It was anticipated that subjects who had solved a P-item in a non-realistic manner themselves during Test 1, would sometimes notice the problematic nature of that item (and of their non-realistic response to it) after being confronted with the realistic response alternative during Test 2, and would therefore score the RA and the NA in a manner that does not match with their own NR given during Test 1. This prediction was based on the findings from another investigation from the same research group (reported in Chapter 3, pp. 38–42) in which they analyzed upper elementary school children's zone of proximal development with respect to realistic mathematical modeling. Indeed, in that study it was found that a small but significant number of pupils who spontaneously answered a P-item with an NR replaced it by an RR after being confronted with the realistic response of a fictitious classmate.

A second possible source of mismatches between the student-teachers' own solutions of the P-items during Test 1 and their evaluations of the realistic and the non-realistic response alternatives during Test 2, relates to their pedagogical content knowledge and beliefs about the role of real-world knowledge in school arithmetic word problem solving. For example, some subjects who had themselves demonstrated awareness of the problematic nature of a P-item during Test 1, may nevertheless give scores for the RA and for the NA reflecting little or no appreciation for pupil answers based on realistic context-based considerations, because they believe that the activation of such considerations should not be stimulated but rather discouraged in elementary school mathematics. It was hoped that a detailed analysis of these mismatches – and especially of the student-teachers' written explanations for them – would yield interesting information concerning their pedagogical content knowledge and beliefs about the importance of real-world knowledge and realistic considerations in school arithmetic word problem solving.

Results

Test 1
As expected, the student-teachers demonstrated a strong overall tendency to exclude real-world knowledge and realistic considerations when confronted with the problematic word problems. Indeed, only 48% of all reactions to the seven P-items from Test 1 could be considered as realistic (RRs). This percentage is considerably higher than in the studies with upper elementary and lower secondary school pupils reported in Chapter 2 in which overall percentages of

Table 5.1   Percentages of Realistic Reactions (RRs) of the Student-Teachers to the Seven
P-items from Test 1 (Verschaffel et al., 1997).*

Problem	% RRs
Buses	90
School	48
Runner	31
Flask	39
Rope	37
Planks	64
Friends	29
Total	48

*   For the full text of the P-items mentioned in the first column, see Table 2.2. For a
detailed description of the criteria for scoring a reaction as NR or RR, see
Verschaffel et al. (1994).

RRs between 15% and 20% were observed (see Table 2.5). Nevertheless, it
remains remarkably low, as it implies that in more than half of the cases, the
student-teachers reacted to the P-items from Test 1 without any serious consideration for the realities of the context involved in the problem.

There was a significant difference in the overall number of RRs between the
first-year and the third-year student-teachers in favor of the latter group (t-test,
two-tailed, t = 3.40, p < .01). However, the overall percentage of RRs remained
low in both groups, namely 45% and 54% for the first-year and the third-year
students, respectively. Interestingly, the size of the difference between the two
years was dissimilar for the three teacher training institutes involved in the study,
suggesting that student-teachers' disposition toward realistic modeling of arithmetic word problems is at least partially influenced by the courses on mathematics education received during their pre-service training.

As in the previous studies with upper elementary and lower secondary grade
students, certain P-items elicited markedly more RRs than others. Table 5.1 lists
the percentages of RRs for each of the 7 P-items. As shown in this table, the
buses item yielded again the largest number of RRs, namely 90%. Just like
school children, student-teachers seem more inclined to activate relevant real-
world knowledge when dealing with the interpretation of the outcome of a
DWR problem, than when they are confronted with P-items involving other
kinds of realistic modeling difficulties such as the interpretation of additive situations involving sets with joint elements (as in the friends item) or situations
involving the illusion of linear proportionality (as in the flask item).

Test 2
Student-teachers' lack of disposition towards realistic modeling was also
revealed by their evaluations of the realistic answer (RA) and the non-realistic
answer (NA) on the same 7 P-items during Test 2 (see Table 5.2). Only in 47%

Table 5.2 Percentages of 1-, ½-, and 0-Scores for the Realistic Answer and the Non-Realistic Answer on the Seven P-items from Test 2 for the First-year and the Third-year Student-Teachers (Verschaffel et al., 1997).

Year	Realistic answer			Non-realistic answer		
	1	½	0	1	½	0
1	44	7	49	58	25	17
3	53	5	42	52	26	22
Total	47	6	47	56	26	18

of the cases did the RA receive a score of 1, while 6% received a score of ½ and the remaining 47% were scored as 0. On the other hand, the NA was scored as 1 and as ½ in 56% and 26% of the cases, respectively, with only 18% of the NAs receiving a score of 0. Thus, student-teachers' overall evaluation of the non-realistic answers to the P-items was considerably more positive than for the realistic answers based on context-based considerations.

There was, again, a significant difference between the first-year and third-year student-teachers. Third-year students gave significantly more scores of 1 (t-test, two-tailed, $t = 3.29$, $p < .01$) and fewer scores of 0 (t-test, two-tailed, $t = 2.63$, $p < .01$) for the RAs than first-year students. Correspondingly, third-year students produced significantly fewer scores of 1 (t-test, two-tailed, $t = 2.30$, $p < .05$) and more scores of 0 (t-test, $t = 2.33$, $p < .05$) for the NAs than first-year students. As for Test 1, the size of the difference between first-year and third-year students varied noticeably among the three teacher-training institutes.

As in Test 1, certain P-items (most notably the buses item) elicited much better scores for the RA and much lower scores for the NA than others (for more details see Verschaffel et al., 1997).

Relationship between Test 1 and Test 2
So far the results have been discussed for Test 1 and Test 2 separately. The final part of the analysis involved investigating to what extent student-teachers' evaluations of the NAs and the RAs during Test 2 matched their own performances during Test 1, by analyzing the scores for the RA and the NA following the 52% non-realistic reactions (NRs) and the 48% realistic reactions (RRs) on Test 1 separately.

Table 5.3 presents the distribution of the different combinations of RA scorings (1, ½ or 0) and NA scorings (1, ½ or 0) given on Test 2 across all 7 P-items for the cases where the corresponding item was answered by the same subject with an NR on Test 1 (52% of all cases), as well as the distribution of the scorings for the RA and the NA over the three scores (1, ½ and 0).

As expected, a strong relationship was found between the NRs on a P-item during Test 1, and the evaluations of the RA and the NA for that item during Test 2. In 89% of all cases for which an NR was given to a P-item during Test

Table 5.3   Percentages of Combinations of Scores (1, ½, or 0) for the Realistic Answer
(RA) and for the Non-Realistic Answer (NA) over the Seven P-items of Test
2 for the Student-Teachers who Themselves Produced a Non-Realistic
Reaction (NR) on the Corresponding Item in Test 1 (Verschaffel et al., 1997).

Score for NA	Score for RA			Total
	1	½	0	
1	3	7	79	89
½	4	0	3	7
0	3	0	1	4
Total	10	7	83	100

1, the NA to that item was given a 1-score in Test 2. Correspondingly, 83% of
the NRs during Test 1 were followed by a score of 0 for the RA in Test 2. The
combination of a score of 1 for the NA and a score of 0 for the RA occurred in
no less than 79% of all cases for which a P-item from Test 1 was answered with
a NR. Apparently, the NA was scored with 1 because this response corre-
sponded to the (non-realistic) answer the student-teachers had themselves given
on this item during Test 1, and they scored the RA with 0 because they could
not understand, and, therefore, appreciate the context-based considerations
underlying this answer.

10% of the NRs to a P-item during Test 1 were followed by a score of 1 for
the RA during Test 2. This suggests that in those cases the confrontation with
the RA during Test 2 had functioned as a scaffold toward (more) realistic modeling. However, the finding that only 10% of the scorings following a NR yielded evidence for the scaffolding effect of the confrontation with the RA can be
interpreted as additional evidence for the strength of the tendency among student-teachers not to take realistic considerations into account, and to resist
being influenced in that direction. Interestingly, these 10% scorings showing
insight into the appropriateness of the RA during Test 2 – as evidenced by the
score of 1 for the realistic response alternative – were accompanied by a diversity of appreciations of the NA (0, ½ and 1). An explanation for these scoring
patterns will be given below (when we discuss similar scoring patterns following an RR during Test 1).

Table 5.4 presents the distribution of the different combinations of RA scorings (1, ½ or 0) and NA scorings (1, ½ or 0) given on Test 2 across all 7 P-items
for the cases where the corresponding item was answered by the same subject
with an RR on Test 1 (48% of all cases), as well as the distribution of the scorings for the RA and the NA over the three scores (1, ½ and 0).

As shown in this table, the congruence between the RRs on Test 1 and the
scorings of the RA and the NA during Test 2 was less straightforward than for
the NRs. The evaluations of the RA were generally in line with the reactions on
Test 1. Indeed, 85% of the RRs on Test 1 were followed by a score of 1 for the
RA on Test 2. However, the scorings for the NA were rather surprising – only

Table 5.4   Percentages of Combinations of Scores (1, ½, or 0) for the Realistic Answer
(RA) and for the Non-Realistic Answer (NA) over the Seven P-items of Test
2 for the Student-Teachers who Themselves Produced a Realistic Reaction
(RR) on the Corresponding Item in Test 1 (Verschaffel et al., 1997).

	Score for RA			
Score for NA	1	½	0	Total
1	10	3	8	21
½	42	2	2	46
0	33	0	0	33
Total	85	5	10	100

33% of the cases where a subject reacted in a realistic way to a P-item during
Test 1 were following by scoring the NA with a 0 during Test 2 (almost always
in combination with a 1 for the RA). A closer look at Table 5.4 reveals that one
scoring combination occurred even more frequently than the combination of a
1 for the RA with a 0 for the NA, since in 42% of the cases for which an RR
was given to a P-item from Test 1, the 1 score for the RA on that item was
accompanied by a ½ for the NA. In addition, the combination "1 for RA and 1
for NA" also occurred in a substantial number of cases (10%). These results
indicate that in many instances where student-teachers reacted themselves to a
P-item in a realistic manner, they were nevertheless quite understanding and tol-
erant towards elementary school pupils who interpreted and solved these P-
items without seriously taking into account the relevant knowledge about the
context called up by the problem statement. According to their written explanations in the comments box, they thought that it would be unfair to punish a
fifth-grader for solving the P-item in a stereotyped, non-realistic manner. This is
illustrated by the following comment accompanying the scoring combination "1
for RA and 1 for NA" with respect to the runner item: "I scored alternative D
(the RA, which was "It is impossible to answer precisely what John's best time
on 1 kilometer will be") with 1 because the pupil who gave this answer knows
that is not realistic to assume that John will be able to run at his record speed
for 1 kilometer. But I also gave 1 for alternative A (the NA, which was "17 ×
10 = 170. John's best time to run 1 kilometer is 170 seconds") because from a
purely computational point of view this is the correct answer."
    Interestingly, a considerable percentage of scoring combinations following an
RR during Test 1 involved a score of ½ (5%) or even a 0 (10%) for the RA. A
qualitative analysis of the written protocols accompanying these unexpectedly
low scores for the RA (taking into account that the student-teacher had produced a RR on this item during Test 1), revealed that the RA was appreciated
so moderately because of its vagueness – to deserve a better score, the RA should
have been more precisely formulated and/or better motivated. This is illustrated in the following illustrative comment for a score of ½ for the RA on the rope
item ("It is impossible to know how many pieces of rope you will need"): "In

fact the boy who has answered in this way is right because you do not know
how much you will lose for making the ties, but he should have explained this
in his answer".

Conclusion
The previous chapters have provided ample evidence for a phenomenon whereby upper elementary and lower secondary students solving word problems in
school often produce answers without regard for realistic constraints. The study
of Verschaffel et al. (1997) with student-teachers provides some insight into one
of the instructional factors considered responsible for the development of this
tendency among children, namely the teachers' own conceptions and beliefs
about the importance of real-world knowledge in arithmetic word problem solving. While this study convincingly demonstrates that many future teachers have
knowledge and beliefs about teaching and learning arithmetic word problems
that are problematical from a realistic point of view, it does, of course, still not
yield direct evidence that these teachers' conceptions and beliefs are responsible
for children's strong tendency to exclude real-world knowledge from their problem solving endeavors. However, based on the recent literature on the relationship between teachers' mathematical beliefs and behavior in the mathematics
lessons, on the one hand, and students' mathematics learning, on the other hand
(De Corte et al, 1996; Fennema & Loef, 1992; Thompson, 1992), there is good
reason to assume that teachers' cognitions and beliefs about the role of real-
world knowledge in the interpretation and solution of school arithmetic word
problems may have strong impact on their actual teaching behavior and, consequently, on their students' learning processes and outcomes.

Before closing the discussion of this study with student-teachers, we want to
stress that – analogously to how we interpreted upper elementary and lower secondary school pupils' non-realistic answers on the P-items – we do not want to
blame these future teachers for their "irrational" or "mindless" behavior and that
of their pupils. As with the pupils, the (student-)teachers in the study of
Verschaffel et al. (1997) were (sometimes deliberately, but mostly unconscious-
ly) behaving "rationally" in accordance with the requirements they have to
observe as partners of the didactical contract for arithmetic word problem solving. Nevertheless, this does not keep us from being concerned about the apparent lack of awareness of the (complexity of the) "problematique" of
mathematical modeling among the majority of (future) teachers, and from making a plea to pay more attention to this modeling perspective in teacher training.

Summary and discussion

In this chapter, we have reviewed the research evidence documenting possible
flaws in different aspects of instructional environments that may account for the
development of students' tendency to solve word problems in a non-realistic
way. Flaws have been revealed not only in the nature of the problems with

which the students are typically confronted in their mathematics textbooks and assessments, but also in the way these problems (and students' responses to them) are considered and treated by teachers. Both aspects of the instructional environment may jointly sanction students' tendency towards an absence of realistic modeling. Taking all this into account, it does not make sense to describe the problem-solving behavior of the students and of their (future) teachers, as documented in the previous chapters, as ugly or foolish or blind. On the contrary, as a result of schooling, these students' (and student-teachers') behavior is pragmatically functional, as the absence of sense-making is based on "a mental representation not only of the specific task (= problem model) but also of the socio-contextual situation in which they are (= social context model)" (Reusser & Stebler, 1997a, p. 325). For the subjects in these settings, to neglect "realistic" interpretations of word problems is functional, as it leads to right and expected solutions (from the point of view of the students solving the problems) and to appropriate and expected feedback (from the point of view of the teachers). The strategy, thus, has its rational core in a socio-cognitive setting of school mathematics. So, why should students abandon beliefs and tactics that apparently proved so successful in the past? As long as both the stereotyped nature of the problems and the impoverished settings in which they are presented impair the quality of mathematics learning in classroom settings, we can hardly blame students (and student-teachers) for their mathematical beliefs and behavior that can be categorized as "non-realistic" in the sense the term has been used throughout the book, but not in the sense of "being adapted to the realities of the instructional or evaluative setting" (Reusser & Stebler, 1997a).

We acknowledge that the research evidence for the direct effect of instructional and assessment practices on students' beliefs about and strategies for solving mathematics application problems at school reported in this chapter is somewhat impressionistic, and therefore not very conclusive. However, we are not the only painters of this impressionistic picture. It is, rather, a patchwork consisting of the impressions of many scholars who have been looking over the past few years from a similar theoretical perspective at the current instructional and assessment practice with respect to mathematical modeling.

We are also aware that our description of the current practice of word problem solving in school may be oversimplified and too negative. There is no doubt that the ongoing worldwide reform in the domain of mathematics education – as envisioned in documents such as the *Curriculum and Evaluation Standards for School Mathematics* in the United States (National Council of Teachers of Mathematics, 1989), the United Kingdom's *Mathematics Counts* (Cockroft, 1982), the Australian *National Statement on Mathematics for Australian Schools* (Australian Education Council, 1990), the Dutch *Proeve van een Nationaal Programma voor het Reken/wiskundeonderwijs* (Treffers & De Moor, 1990) and the *Eindtermen en Ontwikkelingsdoelen: Wiskunde* in Flanders (Ministerie van de Vlaamse Gemeenschap, 1997) – may lead to a change in teachers' and students' attitudes towards the role of reality-based thinking in mathematics teaching and learning (De Corte et al., 1996;

Verschaffel & De Corte, 1996). Yet the dynamics of the ongoing worldwide reform and even the presence in the literature of a growing number of good illustrations of authentic mathematical activities and assessment materials do not necessarily imply that the picture of the daily reality of teaching and learning arithmetic word problem solving presented in this chapter is no longer generally valid. Indeed, there is evidence that the manifest shift at the rhetorical level of official discourse about mathematics teaching and assessment – as could be illustrated by numerous quotations from the above-mentioned reform documents – has not yet led to corresponding changes in the minds and the behavior of the majority of the practitioners (Cooper, 1992, 1994; Gravemeijer, 1994).

6

Beyond Ascertaining Studies: Applying the Modeling Perspective

Invent stories belonging to the numerical problem "6394 divided by 12" such that the result is, respectively:

- *532*
- *533*
- *532 remainder 10*
- *532.83 remainder 4*
- *532.83333 and*
- *about 530.*

(Streefland, 1988, p. 81)

In Chapter 5 we considered the key factors in the instructional environment – the textbooks, the assessment instruments, and, last but not least, the teachers – that we believe underlie the general phenomenon of non-realistic mathematical modeling documented in Part 1 of the book. In this chapter, we move from ascertaining studies to intervention studies, that is to say from studies that primarily document the state of affairs as it exists, to studies that seek to intervene in, and thereby improve, the state of affairs. The first two examples were carried out in one of our research centers and are reported in some detail; for the other two examples, we summarize the work of other researchers and mathematics educators. The first three examples are design experiments with upper elementary school pupils all of which had as at least one of their major goals the development of a disposition towards (more) realistic mathematical modeling and problem solving. While the first one (Verschaffel & De Corte, 1997b) is directly related to the kind of problems we focused on in previous chapters, the second (Verschaffel, De Corte, Lasure, et al., 1999) and the third (Cognition and Technology Group at Vanderbilt, 1997) have a broader aim and scope. We end the chapter with illustrative examples of attempts to improve assessment and an example of a study that provides some direct evidence of the effect that assessment has on teaching through monitoring the effects of radical changes in assessment methods in the state of Victoria in Australia (Clarke & Stephens, 1996).

Teaching realistic mathematical modeling: An exploratory teaching experiment

Starting from the findings of their ascertaining studies reported in previous chapters, Verschaffel and De Corte (1997b) set up a small-scale teaching experiment in which they tried to change pupils' conceptions of the role of real-world knowledge in mathematical modeling and problem solving, and to develop in them a disposition towards (more) realistic mathematical modeling. This was attempted by immersing them (albeit for a short time) into a different classroom culture in which word problems were explicitly conceived and used as exercises in realistic mathematical modeling.

Design
Three classes from the same school participated in the experiment, comprising one experimental (E) class of 19 fifth-grade children, and two control classes (C1 and C2) of 18 and 17 sixth-grade children, respectively. The pupils from the E-class participated in an experimental program on realistic modeling consisting of five teaching/learning units (TLUs) of about 2 ½ hours each, spread over a period of 2–3 weeks. The major characteristics of the program may be described as follows.

First, the impoverished and stereotyped diet of standard word problems offered in traditional mathematics classrooms was replaced by a set of more realistic non-routine problem situations that were specifically designed to stimulate pupils to pay attention to the complexities involved in realistic mathematical modeling, and to distinguishing between realistic and stereotyped solutions of mathematical applications.

Each TLU focused on one prototypical problem of mathematical modeling.

- The topic of the first TLU was making appropriate use of real-world knowledge and realistic considerations when interpreting the outcome of a division problem involving a remainder. The opening problem involved a story about a regiment of 300 soldiers doing several military activities. Each part of the story was accompanied by a question that always had at its core the same arithmetic operation (namely 300 ÷ 8) but required each time a different interpretation to provide an answer appropriate to the situation described, these answers being, respectively, 38, 37, 37.5 and 37 remainder 4 (cf. the example from Streefland cited at the start of this chapter).

- The theme of the second TLU was the union or separation of sets with joint elements. The opening problem was about a boy who already possessed a given number of comic strips from the popular series "Suske and Wiske" and got a package of second-hand albums of Suske and Wiske from his older cousins (who had lost their interest in these comics). The pupils had to determine how many albums of Suske and Wiske there were still missing in the boy's collection after getting this present, given the total number of available albums in this series (236), the number of albums in the boy's collection before the gift (96), and the number of albums in the package he got from his cousins (45, some of which were already in the boy's collection).
- The third TLU focused on problem situations in which it is not immediately clear whether the result of adding or subtracting two given numbers yields the appropriate answer or an answer that is 1 more or 1 less than the correct one. In the opening problem, pupils were given the number of the first and the last ticket sold at the cash desk of a swimming pool on a particular day (ticket numbers 524 and 616, respectively), and they had to work out how many tickets had been sold that day.
- The fourth TLU started with a problem about a boy who wanted to make a swing and had to decide about the amount of rope needed for fastening the swing to a horizontal branch of a big old tree at a height of 5 meters. In that session, pupils had to become aware that in many application problems one has to take into account several relevant elements that are not explicitly nor immediately given in the problem statement but that belong to one's common-sense knowledge base.
- The fifth, and final, TLU dealt with the principle of proportionality and, more particularly, with how to discriminate between cases where solutions based on direct proportional reasoning are and are not appropriate. The starting problem was about a young athlete whose best time to run 100 meters was given, and pupils were asked to predict his best time on the 400 meters.

A second major characteristic was that not only the problems but also the teaching methods differed considerably from traditional mathematics classroom practice. The opening problem in each TLU was tackled in six mixed-ability groups of 3–4 pupils. This group assignment was followed by a whole-class discussion, in which the answers, the solution processes, and possible additional comments of the different groups were compared. Then, each group was given a set of four or five new problems, some with and some without the same underlying modeling difficulty as the opening problem. This group assignment was again followed by a whole-class discussion. Finally, each pupil was individually administered one problem that involved once again the modeling difficulty associated with the TLU. Pupils' reactions to this individual assignment were also discussed afterwards during a whole-class discussion. By way of illustration,

---

TLU1 Opening problem: A soldier's day

300 soldiers must be transported by jeep to their training site. Each jeep can hold 8 soldiers. How many jeeps are needed?

At the training site, the soldiers are brought to a hangar. This hangar is filled with a large number of heavy crates that need to be moved to another hangar. These crates are so heavy that it requires 8 men to carry each one. How many crates can be moved at the same time by the 300 soldiers?

Back in the barracks, all the soldiers are very hungry. The cook has prepared 300 litres of hotpot. He needed 8 big pots of the same size, all completely full, to make the hotpot. How many litres of hotpot does one pot contain?

In the evening, the soldiers have to participate in a military parade. They have to form rows of 8. How many soldiers are left over after forming as many rows as possible?

Fig. 6.1   Worksheet 1 of TLU1 from the study of Verschaffel and De Corte (1997b).

TLU1 Opening problem: A soldier's day (response sheet)

1.   How many jeeps are needed?

Calculation:

Answer:

2.   How many crates can be moved to the other hangar at the same time?

Calculation:

Answer:

3.   How many litres of hotpot does one pot contain?

Calculation:

Answer:

4.   How many soldiers are left after making as many rows as possible?

Calculation:

Answer:

5.   Did you notice anything special when answering these four questions?

Fig. 6.2   Worksheet 2 of TLU1 from the study of Verschaffel and De Corte (1997b).

Figures 6.1–6.4 show the worksheets used in the various instructional phases of the first TLU.

The third major characteristic of the program was that a carefully planned attempt was made to establish a different classroom culture by explicitly negotiating new socio-mathematical norms about the role of the teacher and the students in the classroom, and about what counted as a good mathematical problem, a good solution procedure, and a good response (see Cobb et al., 1992;

TLU1 Application problems

1. Grandfather gives his grandchildren – Paul, Sven, Kurt, Bram, Sarah, and Charlotte – a box containing 75 balloons, which they share equally. How many balloons does each grandchild get?

Calculation:

Answer:

2. Eighty parents participate at a parents' association meeting. Six parents can sit at each table. How many tables are needed?

Calculation:

Answer:

3. Two hundred and fifty eggs are packed in 12 boxes. How many complete boxes can be filled and how many eggs remain in the last incomplete box?

Calculation:

Answer:

4. At the market father bought a box of apples. He paid 300 francs. At home, he weighs the apples and finds that he has exactly 24 kg. apples. "This means that I paid . . . francs per kilogram", he said.

Calculation:

Answer:

Fig. 6.3   Worksheet 3 of TLU1 from the study of Verschaffel and De Corte (1997b).

TLU1 Individual task

Invent stories belong to the numerical problem $100 \div 8 = \ldots$ such that the result is, respectively, 12, 13, and 12.5

Good luck!

Story 1 (the answer must be 12)

Story 2 (the answer must be 13)

Story 3 (the answer must be 12.5)

Fig. 6.4   Worksheet 4 of TLU1 from the study of Verschaffel and De Corte (1997b).

Gravemeijer, 1994, 1997; Schoenfeld, 1988). Examples of how this was established are given below (p. 98).

During the experiment, the pupils from the two control classes followed the regular mathematics curriculum. Before the experimental program was applied in the E-class, the three groups were given the same pre-test, which contained 10 P-items (2 items per TLU) and five standard word problems which functioned as buffer items. One problem in each pair of P-items was similar to an

Table 6.1   The Ten P-items from the Pre-Test Used in the Study of Verschaffel and De Corte (1997b).*

1A   1180 supporters must be bused to the soccer stadium. Each bus can hold 48 supporters. How many buses are needed?

1B   228 tourists want to enjoy a panoramic view from the top of a high building. In the building there is only one elevator. The maximum capacity of the elevator is 24 persons. How many times must the elevator ascend to get all tourists to the top of the building?

2A   At the end of the school year, 66 school children try to obtain their swimming diploma. To get this diploma one has to succeed in two tests – swimming 100 meters breaststroke in 2 minutes and treading water for one minute. 13 children do not succeed in the first test and 11 fail on the second one. How many children get their diploma?

2B   Carl and Georges are classmates. Carl has 9 friends he wants to invite for his birthday party, and Georges 12. Because Carl and Georges have the same birthday, they decide to give a joint party. They invite all their friends. All friends are present. How many friends are there at the party?

3A   Some time ago the school organized a farewell party for its principal. He was the school's principal from January 1, 1959 until December 31, 1993. For how many years was he the principal of that school?

3B   This year the annual rock festival Torhout/Werchter was held for the 15th time. In what year was this festival held for the first time?

4A   A man wants to have a rope long enough to stretch between two poles 12 meters apart, but he has only pieces of rope 1.5 meters long. How many of these pieces would he need to tie together to stretch between the poles?

4B   Steve has bought 4 planks of 2.5 meters each. How many planks of 1 meter can he saw out of these planks?

5A   Sven's best time to swim 50 meters breaststroke is 54 seconds. How long will it take him to swim 200 meters breaststroke?

5B   This flask is being filled from a tap at a constant rate. If the depth of the water is 4 cm after 10 seconds, how deep will it be after 30 seconds?

---

*   The first (A) problem of each pair of P-items is the learning item, whereas the second (B) problem is the near-transfer item.

item from the corresponding session of the experimental program (each such problem will be referred to as a learning item), while in the other problem the same underlying mathematical modeling difficulty had to be handled in a different problem context (referred to as a near-transfer item) (see Table 6.1).

At the end of the experimental course, a parallel version of the pre-test was administered in all three classes as a post-test. However, in one of the control classes, labeled C1, the post-test was preceded by a 15 minutes introduction in which the pupils were explicitly warned that the test would contain several problems for which routine solutions based on adding, subtracting, multiplying or dividing the given numbers, would be inappropriate. This procedure was, thus, similar to the minimal interventions aimed at alerting students to the problematic aspect of P-items discussed in Chapter 3.

One month after the post-test, a retention test was administered in the E-class consisting again of 10 P-items, half of which were parallel versions of the near-transfer items from the pre-test and post-test (these constitute near-transfer items), whereas the other half were even more remote from those encountered during the experimental program, such as the water item (P3) and the school-distance item (P6) from Table 2.2 (constituting far-transfer items).

Hypotheses and additional research questions

First, in line with the results of the studies of Verschaffel et al. (1994) and Verschaffel, De Corte, and Lasure (1999), it was hypothesized that on the pre-test the pupils would demonstrate a strong tendency to exclude real-world knowledge when confronted with the problematic versions of the problems. Therefore, it was predicted that on the P-items of the pre-test the pupils of the three classes (E, C1 and C2) would produce a percentage of RRs that would not differ significantly from the 15–20% RRs obtained in these previous studies (see Chapter 2).

A second hypothesis was that the experimental program would result in a disposition toward (more) realistic modeling and interpreting of arithmetic word problems. Therefore, a significant increase in the number of RRs on the P-items from pre-test to post-test for the E-group was predicted. For the two other groups no significant increase from pre-test to post-test was expected.

Third, a positive transfer effect of the experimental program was hypothesized. Therefore, it was predicted that the increase in the percentage of RRs in the E-group from pre-test to post-test, would be significant not only for the five items that were similar to those from the experimental program (the learning items), but also for the five near-transfer items.

A fourth hypothesis was that the positive effect of the experimental program would be lasting. This led to the prediction that there would be no significant difference between the percentage of RRs of the E-group for the P-items on the retention test and for the parallel problems on the post-test.

Besides testing these four hypotheses on the basis of pupils' overall performances on the different tests, two additional research questions were addressed:

- Was the experimental program equally effective for pupils with different levels of mathematical ability (i.e., for mathematically strong, average, and weak pupils) in the E-class?
- Was the program equally effective for the five types of mathematical modeling difficulties addressed in the respective TLUs of the program, and in the test?

Results

To evaluate the effect of the experimental program, an Analysis of Variance was performed with group (E versus C1 versus C2), time (pre-test versus post-test) and problem type (learning items versus near-transfer items) as the independent variables and the proportion of RRs on the P-items (scored in the same way as in the previously described ascertaining studies) as the dependent variable.

As predicted in the first hypothesis, pupils from all three classes demonstrated a strong tendency to exclude real-world knowledge and realistic considerations from their problem solutions during the pre-test. Overall, the percentage of RRs for the 10 P-items of the pre-test was only 15%, a percentage comparable to those obtained in the authors' ascertaining studies described in Chapter 2. The fifth-graders from the E-class produced fewer RRs than the sixth-graders from the C1- and the C2-classes, the percentages being 7%, 20%, and 18%, respectively. However, using Tukey a posteriori tests the differences were not found to be statistically significant.

To test the second hypothesis, it was analyzed whether there was an interaction effect between the independent variables Group (E, C1, and C2) and Time (pre-test versus post-test). The Analysis of Variance revealed that this Group × Time interaction was indeed statistically significant (p < .0001). Tukey a posteriori tests showed that there was a significant increase in the number of RRs from pre-test to post-test for the E-group (from 7% RRs to 51%). By contrast, in the two control classes the increase in the number of RRs from pre-test to post-test was non-significant, namely from 20% to 34% for C1, and from 18% to 23% for C2 (see Table 6.2). Although the increase from pre-test to post-test in the performance of the experimental group was very substantial (compared to the two control groups), it should be pointed out that in the post-test still only about half of the P-items were solved in a realistic way.

The third hypothesis, about the transfer effect, was tested by analyzing the results of the E-class separately for the five learning items and the five near-transfer items. The lack of a significant Group × Time × Problem Type interaction indicated that the increase in the percentage of RRs in the E-class from pre-test to post-test could not be attributed to a task-specific training effect.

Table 6.2 Numbers and Percentages of Realistic Reactions on the 10 P-items from the Pre-Test and the Post-Test for the Experimental (E) Class and the Two Control Classes (C1 and C2) in the Study of Verschaffel and De Corte (1997b).

Groups	Pre-test		Post-test	
	N	%	N	%
E (n=18*)	13	7	91	51
C1 (n=18)	36	20	61	34
C2 (n=17)	32	19	39	23

* Due to illness one pupil from the E-group could not participate in the post-test.

Indeed, while the increase in the percentage of RRs in the E-class was larger for the five learning items (from 9% to 60%) than for the five near-transfer items (from 6% to 41%), Tukey a posteriori tests revealed that the increase was statistically significant for both kinds of problem.

Finally, the positive results of the E-class on the retention test supported the fourth hypothesis. The percentage of RRs for the five near-transfer items of the retention test was 40%, which was almost the same as the result for the five parallel near-transfer items from the immediate post-test (41%). Remarkably, the five far-transfer items from the retention test elicited also almost the same mean percentage of RRs, namely 39%. This is considerably higher than the percentages observed in equivalent groups of pupils who solved the same P-items without special training in realistic mathematical modeling (see the studies reported in Chapter 2). This finding provides additional evidence that the effect of the program was not task-specific.

As explained earlier, the experimenters also analyzed pupils' individual responses on the different tests to trace possible differential effects of the program on pupils with varying mathematical ability, as well as on the distinct categories of mathematical modeling difficulties involved in the program.

To get an idea of whether the effectiveness of the program depended on the pupils' mathematical ability the percentages of RRs on the P-items of the pre-test, the post-test, and the retention test were separately computed for the mathematically weak, average, and strong pupils. Placement at the three ability levels was based on pupils' achievement scores for mathematics at the end of grade 4 (which was also used to divide the pupils among the six mixed-ability groups for the group assignments). As shown in Table 6.3 the program was especially effective for the strongest pupils in the E-class.

Similarly, the number of RRs in the E-class on the pre-test, the post-test, and the retention test was computed separately for each of the five types of P-items, each type corresponding to the topic of one TLU (see Table 6.4). The data show that the program was more effective for some kinds of mathematical modeling

Table 6.3   Numbers (N) and Percentages (%) of Realistic Reactions on the P-items from the Pre-Test, Post-Test, and Retention Test* for the Mathematically Weak, Average, and Strong Pupils of the Experimental Class in the Study of Verschaffel and De Corte (1997b).

Ability level	Pre-test		Post-test		Retention test	
	N	%	N	%	N	%
Weak (n=6)	4	7	20	33	14	23
Average (n=6)**	4	7	24	40	20	33
Strong (n=6)	5	8	47	78	37	62

*   The pre-test, post-test, and retention test consisted of 10 P-items each.
**   Due to illness one mathematically average pupil could not participate in the post-test.

Table 6.4   Numbers (N) and Percentages (%) of Realistic Reactions of the Pupils of the
            Experimental Class (n=18) on the P-items from the Pre-Test, Post-Test, and
            Retention Test Corresponding to the Five Distinct Teaching/Learning Units
            (TLUs) Used in the Study of Verschaffel and De Corte (1997b).

TLU	Pre-test		Post-test		Retention test*	
	F	%	F	%	F	%
TLU 1**	8	22	30	83	16	89
TLU 2	0	0	20	56	6	33
TLU 3	5	14	13	36	4	22
TLU 4	0	0	16	44	5	28
TLU 5	0	0	2	33	5	28

*   Only the five near-transfer items of the retention test are included in this table. The
      five far-transfer items were excluded because their underlying mathematical model-
      ing difficulty does not directly correspond with the topic of one of the TLUs.
      Accordingly, the retention test contained only one P-item per TLU, while in the pre-
      test and the post-test there were two P-items per TLU.
** For details of the topics of the five TLUs see the text.

difficulties than for others. The program's success was most visible for the items
about the interpretation of the outcome of a division with a remainder (TLU1).
The smallest increase in the percentage of RRs was observed for (a) problems
where it was unclear whether an arithmetic operation with the given numbers
would yield the correct answer or an answer that was one more or one less than
the correct one (TLU 3), and (b) those requiring discrimination between situations where direct proportional reasoning was or was not appropriate (TLU 5).

Replication with German pupils
Having ascertained that German students are as "unrealistic" word problem
solvers as students from other countries (see Chapter 2), and having done a
small interventional pilot study with promising results, Renkl (1999) set up a
teaching experiment to replicate also Verschaffel and De Corte's (1997b) finding that it is possible to positively affect pupils' disposition toward (more) mathematical modeling of school word problems by means of an intervention that
tries to reduce the problem of compartmentalization between students' school
and everyday knowledge.

Two groups of fourth-graders participated in the experiment, which took
place at the end of the school year. Pupils from an experimental class (n = 23)
were involved in an intervention which took three lessons of 45 minutes and in
which pupils were confronted with four problematic word problems taken from
the test of Verschaffel et al. (1994), namely the friends item (P1), the buses and
the balloons items (P4 and P7), and the school item (P6) (see Table 2.2). With
respect to each problem children were invited to play "trip-trap" detectives: they
had to detect and explain the problematicity of these word problems. This

investigational work was followed by a whole-class discussion. During these activities pupils represented the problems in different modes (i.e., enactive, visual, and symbolic). Meanwhile the pupils of a parallel control class (n = 23) followed the regular mathematics program.

Before and after the intervention both groups received a test consisting of eight P-items, half of which were near-transfer items (= items with similar mathematical modeling traps as the four P-items they had encountered during the lessons, but embedded in different cover stories) and half of which were far-transfer items (= P-items containing traps that were different from those encountered during the lessons). For instance, for the friends item (P1) handled during the intervention, the following item was used as a near-transfer pre-test and post-test item: "During the soccer world championship, three coaches are interviewed. They have to name the 11 world's best players. The reporter records the names. How many different players will he have on his list at the end?". The following problems from the test of Verschaffel et al. (1994) were used as far-transfer items: the runner item (P5), the planks item (P2), the water item (P3), and the rope item (P9) (see Table 2.2).

An analysis of the answers of the pupils from the experimental and control class on the near-transfer and far-transfer items of the pre-test and the post-test revealed a significant interaction between group and time of test. Whereas the control class showed no progression from pre-test to post-test either for the near-transfer items (from 30% RRs to 34% RRs) or for the far-transfer items (from 1% to 2%), the experimental class demonstrated a significant progression not only on the near-transfer items (from 35% to 57%) but also on the far-transfer items (from 3% to 33%). The remarkably high difference in percentages of RRs between the near-transfer and the far-transfer items is due to the fact that half of the four near-transfer items were problems about the interpretation of the outcome of a division with a remainder, a type of mathematical modeling difficulty that has been found as considerably more easy to overcome in all available studies (see Chapter 2). Finally, an analysis of pupils' individual responses revealed that ability played an important role in the number of RRs produced.

Conclusion
Particularly bearing in mind the relatively short-term nature of the interventions, the results of Verschaffel and De Corte (1997b) and Renkl (1999) warrant a positive conclusion about the feasibility of altering the conception about the role of real-world knowledge in mathematical modeling and problem solving in children of the upper elementary school, and of developing in them a disposition toward (more) realistic mathematical modeling of school word problems. Nevertheless, some caution is in order. First, the reported positive results should be qualified by acknowledging some methodological weaknesses of these exploratory teaching experiments, such as the small size of the experimental and control groups and the absence of a retention test (in the control group). Second, it should be remembered that after the intervention the

overall percentage of RRs of the pupils of the E-classes were still relatively low, indicating that the intervention did not completely solve the problem. Moreover, analyses of the individual test scores revealed great interindividual differences with respect to the number of RRs produced before the beginning of the intervention and to the pupils' susceptibility to that intervention; the increase in the number of RRs was also considerably greater for some types of P-items than for others. A final critical comment is that in neither of the two studies was the teaching done by the regular classroom teacher, implying that the studies lack ecological validity.

Learning to solve mathematical application problems: A design experiment

Background
Taking into account the results and conclusions of the study reviewed in the previous section, as well as those of Lester, Garofalo, and Kroll's (1989) investigation about the role of heuristics and metacognition in mathematical problem solving, and the initial work of the Cognition and Technology Group at Vanderbilt (CTGV) on the Jasper project (which will be described in the next section), Verschaffel, De Corte, Lasure, et al. (1999) set up a design experiment in which a learning environment for teaching and learning how to model and solve mathematical application problems was developed and tested in a number of fifth-grade classes.

Contrary to the earlier investigation, Verschaffel, De Corte, Lasure, et al.'s (1999) learning environment was realized and tested in a typical classroom context, and not in an instructional setting wherein all the teaching was done by one of the researchers. As such, this study has a higher degree of ecological validity than the preceding one, and, therefore, may document better the practical relevance and utility of the underlying research-based ideas about the modeling perspective in mathematical learning and problem solving.

Aims of the learning environment
The aims of Verschaffel, De Corte, Lasure, et al.'s (1999) learning environment were twofold. A first aim was the acquisition by students of an overall strategy for solving mathematics application problems. The overall strategy consists of five stages, involving a set of eight heuristics which are especially useful in the first two stages of that strategy (see Table 6.5). The design of the overall strategy, the selection of the eight heuristics, and their placement in their matching with a particular stage of the five-stage problem-solving model, were based on an extensive review of the available literature on this topic (De Corte et al., 1996; Lester et al., 1989; Polya, 1957; Schoenfeld, 1992; Verschaffel, 1999)

The second aim was the acquisition of a set of appropriate beliefs and positive attitudes with regard to the solution, the learning, and the teaching of mathematical application problems. Examples of such beliefs are that mathematics

Table 6.5   The Competent Problem-Solving Model Underlying Verschaffel, De Corte, Lasure, et al.'s (1999) Learning Environment.

Step 1:	Build a mental representation of the problem Heuristics: Draw a picture Make a list, a scheme, or a table Distinguish relevant from irrelevant data Use your real-world knowledge
Step 2:	Decide how to solve the problem Heuristics: Make a flowchart Guess and check Look for a pattern Simplify the numbers
Step 3:	Execute the necessary calculations
Step 4:	Interpret the outcome and formulate an answer
Step 5:	Evaluate the solution

problems may have more than one correct answer and that solving a mathematics problem may be effortful and take more than just a few minutes.

Main features of the learning environment

The main features of the learning environment, which can be considered as an extension of those underlying the exploratory teaching experiment reported in the previous section, were the following:

- A varied set of carefully designed complex, realistic, challenging and open problems that ask for the application of the intended heuristics and metacognitive skills that constitute the model of skilled problem solving. Some problems were presented in a purely textual format, but others in the form of a story told by the teacher, a newspaper article, a brochure, a comic strip, a table, or a combination of two or more of these presentational formats.
- A series of lesson plans based on a variety of teacher and learner activities. Most of the lessons follow an instructional model similar to the one used in Verschaffel et De Corte's (1997b) exploratory teaching experiment presented above. So each lesson consisted of one or two small-group assignments solved in fixed heterogeneous groups of three to four pupils, each of which was followed by a whole-class discussion, and an individual assignment which was also followed by a whole-class discussion. During all these activities the teacher's role was to encourage and scaffold pupils to engage in, and to reflect upon, the kinds of cognitive and metacognitive activities involved in the model of competent mathematical problem solving. These encouragements and scaffolds were gradually withdrawn as the pupils became more competent, and took more responsibility for their own learning and problem solving.

- Interventions explicitly aimed at the establishment of new socio-mathematical norms, resulting in a classroom climate conducive to the development in pupils of appropriate beliefs about mathematics and mathematics learning and teaching. These norms related to the role of the teacher and the pupils in the classroom (e.g., "don't expect the teacher to decide autonomously which of the generated solutions is the correct or optimal one; this decision will be taken by the whole class as a community of practice after an evaluation of the pros and cons of all candidate solutions"), and about what counted as a good mathematical application problem (e.g., "many problems can be interpreted and solved in different appropriate ways"), a good response (e.g., "sometimes a rough estimate is an even better answer than an exact number"), or a good solution procedure (e.g., "the solution of an expert problem solver will not always consist of a computation or a set of computations; sometimes, it will be a sketch or a diagram").

Content and organization of the learning environment

The learning environment consisted of a series of 20 lessons of 1 to 1½ hours designed by the research team in consultation and co-operation with the regular class teachers. Because the lessons were taught by the class teachers, they

Wim would like to make a swing from a branch of a big old tree. The branch is at a height of 5 meters from the ground. He has already made a suitable wooden seat for his swing. Now Wim is going to buy some rope. How many meters of rope will Wim have to buy?

Fig. 6.5   Starting problem of the lesson about the heuristic "Use your real-world knowledge" (Verschaffel, De Corte, Lasure, et al., 1999).

Every piece of a dominoes game is divided into two sides, each of which has 0, 1, 2, 3, or 4 dots. Some examples are given.

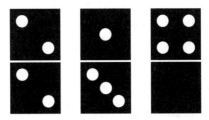

What is the total number of pieces in this game if there is one piece of every possible combination?

What would the total number be if each side of a piece can have a maximum of 6 instead of 4 dots?

Fig. 6.6   Starting problem of the lesson about the heuristic "Look for a pattern" (Verschaffel, De Corte, Lasure, et al., 1999).

were prepared for and supported in implementing the powerful teaching learning environment. With two lesson periods each week, the intervention was spread over about three months. Three major parts can be distinguished in the series of lessons:

- Introduction to the content and organization of the learning environment and reflection upon the difference between a routine task and a real problem (1 lesson).
- Systematic acquisition of the five-step regulatory problem-solving strategy, and the embedded heuristics (15 lessons). Examples of problems used in these lessons are given in Figures 6.5, 6.6, and 6.7.
- Learning to use the competent problem-solving model in a spontaneous and flexible way in so-called project lessons involving more complex application problems the exploration, solution and discussion of which took a whole lesson (4 lessons). (For an example of such complex application problems, see Figure 6.8).

Components of teacher support

Because the lessons were not taught by the researchers but by the regular classroom teachers, these teachers were prepared for and supported in implementing the learning environment. The model of teacher development adopted reflected our views about pupils' learning by emphasizing the creation of a social context wherein teachers and researchers learn from another, rather than a model whereby the researchers transmit knowledge to the teachers (De Corte, in press). Moreover, taking into account that the mathematical teaching/learning process is too complex to be pre-specified and that teaching as problem solving is mediated by teachers' thinking and decision-making, the focus of teacher

A wholesale dealer needs to pack 200 boxes with bulbs in a big wooden case. In the case, there is room for 8 rows with 8 boxes per row and stacked 4 levels high. Is it possible to store the 200 boxes with bulbs in the case?

Below, three incorrect solutions for this problem are given. Find the mistake in each of them.

Solution 1:

$$8 + 8 + 4 = 20$$

No, the man can not store the 200 boxes with bulbs in the case, because the case can only hold 20 boxes.

Solution 2:

$$8 \times 8 \times 4 =$$

$$8 \times 8 = 64 \qquad 64$$

$$\underline{\times\ 4}$$

$$116$$

No, the man can not store the 200 boxes with bulbs in the case, because the case can only hold 116 boxes.

Solution 3:

$$8 \times 8 \times 4 =$$

$$8 \times 8 = 64 \qquad 64$$

$$\underline{\times\ 4}$$

$$256$$

No, because 256 is a lot more than 200.

Fig. 6.7   Starting problem of the lesson about step 5 of the problem-solving model, namely "Evaluate the solution" (Verschaffel, De Corte, Lasure, et al., 1999).

development and support was not on making them perform in a specific way, but on preparing and equipping them to make informed decisions (c.f. Carpenter & Fennema, 1992; Yackel & Cobb, 1996). This form of support involved the following components:

• Provision of a general teacher guide containing an extensive description of the background, the aims, and the overall content and structure of the learning environment, as well as a list of ten general guidelines for the teachers with actions they should take before, during and after the individual or group assignments to increase the "power" of the learning environment (such as "Observe the group work and provide appropriate hints when needed", "Stimulate articulation and reflection" and "Avoid

Pete and Annie are building a miniature town with cardboard. The space between the church and the town hall seems the perfect location for a big parking lot. The available space is in the shape of a square with a side of 50 cm and is surrounded by walls except for the street side. Pete has already made a cardboard square of the appropriate size. What will be the maximum capacity of their parking lot?

1. Fill in the maximum capacity of the parking lot on the banner.
2. Draw on the cardboard square how you can best divide the parking lot into parking spaces.
3. Explain how you came to your plan for the parking lot.

Fig. 6.8   Example of a problem used in one of the project lessons (Verschaffel, De Corte, Lasure, et al., 1999) (Each group received a 50 × 50 cm cardboard and two toy cars to solve this problem).

imposing solutions and solution methods on pupils"). In the teacher guide, each of these ten guidelines was followed by an explanation of its purpose as well as by several worked-out examples to illustrate exactly what it means to act accordingly.

- Provision of a specific teacher guide for each lesson, containing the overall lesson plan with the different mathematics application problems to be addressed in the lesson, as well as the preferable sequence of instructional activities, specific suggestions for appropriate teacher interventions before, during and after the group or individual assignments, examples of anticipated correct and incorrect solutions and solution methods, etc.
- Provision of all the necessary concrete material for the pupils (worksheets, concrete materials related to the application problems, etc.).
- Presence of one member of the research team in each lesson. This researcher did not intervene during the lesson itself, but before and after it (s)he had 5–10 minute discussions, respectively preparatory and evaluative, with the teacher.
- Organization of regular two-hour meetings attended by all members of the research team and by all teachers and headmasters of the experimental schools involved in the experiment. During these meetings, the teachers were invited to comment on first drafts of the general and the specific

teacher guides and to propose and discuss possible extensions and improvements. Besides, teachers and researchers exchanged positive experiences as well as difficulties with the implementation of (certain aspects of) the learning environment and searched for appropriate solutions for these difficulties.

Design of the implementation and evaluation study

The effectiveness of the learning environment was evaluated in a study with a pre-test/post-test/retention test design. Four experimental fifth-grade classes and seven comparable control classes from eleven different elementary schools in Flanders participated in the study.

Three pre-tests were collectively administered in the experimental as well as the control classes:

- A standardized achievement test (SAT) to assess fifth-graders' general mathematical knowledge and skills (Stinissen, Mermans, Tistaert, & Vander Steene, 1985).
- A test consisting of ten difficult non-routine word problems (WPT) that lend themselves to the application of the heuristic and metacognitive skills taught in the experimental program (see Table 6.6 for some examples).
- A questionnaire aimed at assessing pupils' beliefs about, and attitudes towards, (teaching and learning) mathematical application problems (BAQ) consisting of two subscales. One subscale contained seven Likert-like items relating to "pleasure and persistence in solving word problems" (for instance, "I like to solve word problems", "Difficult problems are my favorites"), and the other subscale had 14 items expressing "a problem-and process-oriented view on word problem solving" (for instance, "There is always only one solution to a word problem", "Listening to explanations of alternative solution paths by other pupils is a waste of time").

In addition, pupils' WPT answer sheets for each problem were carefully analyzed to look for evidence of the application of one or more of the heuristics embedded in the problem-solving strategy described in Table 6.5.

Besides these collective pre-tests, three pairs of pupils of equal ability from each experimental class were asked to solve five complex non-routine application problems during a structured interview. The problem-solving processes of these dyads were video-recorded, and afterwards analyzed by means of a schema constructed by the researchers for assessing the intensity and the quality of pupils' use of heuristic and metacognitive strategies.

While the intervention took place in the experimental classes, the control classes followed the regular mathematics program, which also involved a considerable number of lessons in arithmetic word problem solving (for a brief description of the content and organization of these lessons, see Verschaffel, De Corte, Lasure, et al., 1999).

By the end of the intervention, parallel versions of all collective pre-tests (SAT, WPT, and BAQ) were administered in all experimental and control classes. In

Table 6.6   Examples of Items from the Word Problem Pre-Test (WPT) Used in the Study of Verschaffel, De Corte, Lasure, et al. (1999).

Martha is reading a book. Suddenly she finds out that some pages are missing, because page 135 is immediately followed by page 173. How many pages are missing?

Lies has two doll's houses. The square floor of the small doll's house has a side of 40 cm and consists of 16 tiles. The square floor of the large doll's house has a side which is exactly twice the side of the small doll's house. How many tiles are needed for the floor of the large doll's house if the same tiles are used?

Catherine builds little houses with matches. To construct 2 houses, she needs 9 matches. To build a row of 5 houses, she needs 21 matches. How many matches will she need to build a row of 10 houses?

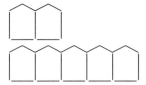

---

addition, the answer sheets of all pupils were again scrutinized for traces of the application of heuristics, and the same pairs of pupils from the experimental classes as were interviewed prior to the intervention were again administered a structured interview involving parallel versions of the five non-routine application problems used during the pre-test.

Three months later a retention test – a parallel version of the collective WPT used as pre-test and post-test – was also administered to all the experimental and control classes.

Finally, to assess the implementation of the learning environment by the teachers of the experimental classes, a sample of four representative lessons was video-taped in each experimental class, and analyzed afterwards to provide an "implementation profile" for each experimental teacher.

Results

The results of this study can be summarized as follows (for more details, see Verschaffel, De Corte, Lasure, et al., 1999).

First, an Analysis of Variance on the results on the WPT using a hierarchical factorial design with a repeated measure with 3 levels for the factor Time (Pre-test, Post-test, Retention test) and with the factors Group (Experimental versus Control Group) and Class (nested in Group) (with four experimental and seven control classes) revealed a statistically significant interaction between Group and Time ($p < .001$). While no statistically significant difference was found between the experimental and control groups on the WPT during the pre-test, the former significantly outperformed the latter during the post-test, and this difference in favor of the experimental group continued to exist on the retention test (see Figure 6.9). This effect had a medium size of .31 (Cohen, 1988). However, it should be acknowledged that in the experimental group pupils' overall

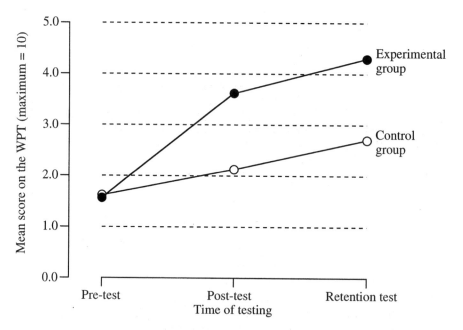

Fig. 6.9   Mean scores of the experimental and the control groups on the three word
problem tests (WPT) (pre-test, post-test, and retention test) (Verschaffel, De
Corte, Lasure, et al., 1999).

performance on the post-test and retention tests was not as high as anticipated
(i.e., the pupils of the experimental classes still produced less than 50% correct
answers on these tests).

   Second, an Analysis of Variance with two levels instead of three for the
repeated measure, Time (Pre-test and Post-test), and BAQ scores as the dependent variable, revealed a significant interaction between Group and Time for
the "Pleasure and persistence" subscale (p < .001) and for the "Problem- and
process-oriented view on word problem solving" subscale (p < .01). The means
of the pupils from the experimental classes changed more in the expected
direction than those of the pupils from the control classes. However, though
statistically significant, the effects were rather small. In both cases an effect size
of only .04 was found.

   Third, a significant interaction between Group and Time was also found for
the total score on the SAT (p < .01). Whereas there was no difference between
the pre-test results on the SAT between the experimental and the control group,
the results on the post-test revealed a significant difference in favor of the former. The effect size was .38 (a medium effect according to Cohen,1988). This
result suggests that the greater attention to mathematical modeling in the experimental classes (at the expense of other subject-matter topics in mathematics)
had no negative side-effect, and even a small positive transfer effect, on pupils'
mathematical knowledge and skills as a whole.

Fourth, an Analysis of Variance with Group, Time and Class (nested in Group) as independent variables, and the number of word problems in the WPT for which (at least) one of the eight heuristics from the competent problem-solving model was visibly used, as the dependent variable, revealed a significant interaction between Group and Time (p < .001). In the experimental group there was a dramatic increase in the use of these heuristics from pre-test to post-test and retention test, whereas in the control classes there was no difference in pupils' use of heuristics between the three testing occasions. Figure 6.10 gives a graphic representation of this interaction effect, which is, according to Cohen (1988), very large (effect size = .76).

Fifth, in line with this result, the videotapes of the problem-solving process-es of the dyads revealed substantial improvement in the intensity and the qual-ity with which the pairs from the experimental group applied certain – but not all – heuristics and metacognitive skills addressed in the learning environment to solve the five complex problems.

Sixth, an Analysis of Variance with the factors Group, Time, and Ability level as independent variables, and the number of correct answers on the WPT as dependent variable, revealed no significant 3-way interaction, suggesting that all three ability groups into which the subjects were categorized (high, medium, and low) contributed significantly to the above-mentioned positive effects in the

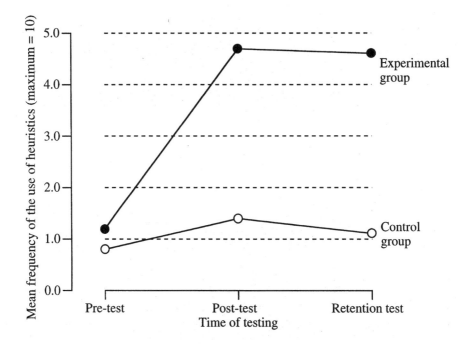

Fig. 6.10   Mean frequency of the use of the heuristics in the experimental and the con-trol groups on the three versions of the word problem test (WPT) (pre-test, post-test, and retention test) (Verschaffel, De Corte, Lasure, et al., 1999).

experimental group, as shown in Figure 6.11. Also, on the other dependent variables – namely pupils' scores on the two BAQ subscales, the SAT, and the frequencies of their use of heuristics on the WPT – no significant 3-way interactions
between Group, Time, and Ability level were found.

Finally, the positive effects of the learning environment were not observed to
the same extent in all four experimental classes. In fact, in one of the four classes there was little or no improvement on most of the process and product measures. The analysis of the videotapes of the lessons in these classes indicated
substantial differences in the extent to which the four experimental teachers had
succeeded in implementing the major aspects of the learning environment.

Conclusion

In this design experiment, a set of carefully designed application problems, a
collection of highly interactive teaching methods, and the introduction of
new socio-mathematical classroom norms were combined in an attempt to
create a substantially modified learning environment that focused on fostering in pupils a mindful approach toward mathematical modeling and problem solving. This learning environment was implemented in four classes of
the fifth grade, and its effects were evaluated in a teaching experiment with
a pre-test / post-test / retention test design involving an experimental group
(four classes) and a comparable control group (seven classes).

According to the results on the various pre-tests and post-test, the learning
environment had a significant positive effect on the development of pupils'
mathematical modeling performance, which did not disappear after the end of

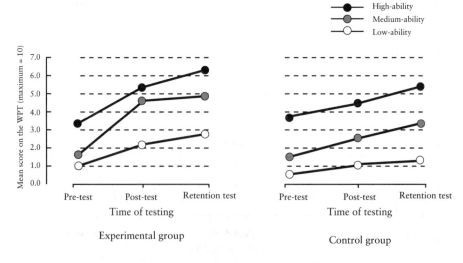

Fig. 6.11   Mean score of the high-, medium-, and low-ability pupils of the experimental and control groups on the three versions of the word problem test (WPT)
(pre-test, post-test, and retention test) (Verschaffel, De Corte, Lasure, et al.,
1999).

the experimental lessons. Moreover, significant positive effects were observed on their mastery of heuristic and metacognitive strategies that are valuable in mathematical modeling, as well as on their beliefs and attitudes about mathematical modeling and problem solving. The results on the standard achievement test showed that the extra attention paid to modeling and problem-solving strategies, beliefs, and attitudes in the experimental classes had no negative influence on the learning results for the other, more traditional aspects of the mathematics curriculum. On the contrary, there was even a small but statistically significant positive transfer effect. Finally, these positive results were not restricted to pupils of high and medium ability in mathematics, but were also found among the low-ability pupils.

Although the overall results are promising, some aspects of the findings warn against over-optimistic conclusions. In this respect, it should be remembered that – except for the drastic increase in the use of heuristics during the collective word-problem test – none of the positive effects of the learning environment on the experimental classes was large in terms of Cohen's (1988) effect size measure. Moreover, the learning environment did not lead to a decrease in the initial differences between pupils of high and low ability on the dependent measures. And finally, on all measures the results of one of the four experimental classes were considerably lower than those of the three other classes.

As a first plausible reason why the obtained positive effects were not greater, it should be borne in mind that the intervention consisted of only 20 lessons given rather independently of the rest of the mathematics curriculum. Presumably, the results would have been better if the available instructional time had been greater, and if the learning environment could have been integrated more fully within the regular mathematics lessons. Second, although the teachers were prepared for their new tasks and roles as much as possible (within the constraints of the project), both the teachers and the researches had the impression that this preparation was not long and intensive enough, taking into account the rather drastic changes in the teachers' pedagogical content knowledge and beliefs and in the teachers' teaching behavior required by the new learning environment. Finally, further elaboration of the three pillars of the learning environment (based on the findings of the current design experiment) should lead to still better results. This holds especially for the third pillar, namely the establishment of a new classroom culture, which was still not addressed in a sufficiently systematic and effective way in the study. With respect to the two other pillars, the authors point out the need to search for a better balance between the requirement that the problems can elicit the intended (meta-)cognitive activities, on the one hand, and the realistic or authentic nature of these problems, on the other. With regard to the instructional techniques, they consider as a major challenge for further research how to organize and support the small-group activities so that all pupils – including the low-ability ones – interact and cooperate in a high-quality task-oriented manner.

Anchoring mathematical problem solving in realistic contexts using new
information technologies: The Jasper series

Background of the Jasper series

As in the two previous design experiments, the Cognition and Technology
Group at Vanderbilt (CTGV) (1997; Van Haneghan, Barron, Young, Williams,
Vye, & Bransford, 1992) started from the critical observation that, as traditionally taught, pupils' experiences with word problem solving consist mainly
of choosing the arithmetic operation with the given numbers in the problem to
figure out the correct answer, without paying attention to the realities of the
context evoked by the problem statement. This is far removed from solving
mathematical application problems in the real world, in which posing and defining problems, model building, planning and decision-making, and interpreting
outcomes – always linked to the real-life context – are major activities.
However, while the tasks used by Verschaffel and De Corte (1997b) and even
by Verschaffel, De Corte, Lasure, et al. (1999) still showed some resemblance
to the kind of word problems used in the traditional word problem solving
lessons, the CTGV (1997) applied videodisc technology to confront pupils with
rich, authentic, and complex problem-solving contexts offering ample opportunities for problem posing, exploration, and discovery.

Although the researchers of the CTGV stress that the application of such
"anchored instruction" does not necessarily require the use of technology, they
nevertheless argue that a technology-based implementation makes the teach-
ing/learning environment more powerful (and in their later work technology
plays an even greater role than in their initial studies, as discussed below).
Accordingly, they have developed videodisc-based complex problem spaces
enabling learners to explore scenarios involving mathematical problems for
extended periods of time and from a diversity of perspectives. However, these
videodisc-based problem spaces are only one of the components that they
believe to be important. Other conditions include (1) the guidance provided by
an expert teacher, who organizes and designs the learning experience, who stim-
ulates cooperative learning and discussion in small groups, and who explicitly
addresses the culture of the classroom, and (2) the availability to children of rich
and realistic sources of information.

The adventures of Jasper Woodburry

One series of videodiscs for mathematics instruction in grade 5 and 6 developed
by the Vanderbilt group is called "The Adventures of Jasper Woodburry". It
contains 12 different dramatically presented adventures covering a variety of
mathematical concepts (distance/time/rate, statistics and probability, geometry,
and algebra). In the initial videodisc of this series, called "Journey to Cedar
Creek", Jasper Woodburry takes a river trip to see an old cabin cruiser he is
considering purchasing. Jasper and the cruiser's owner test-run the cruiser, after
which Jasper decides to purchase the boat. As the boat's running lights are

inoperative, Jasper must determine if he can get the boat to his home dock before sunset. Two major questions that form the basis of Jasper's decision are presented at the end of the disc:

- Does Jasper have enough time to return home before sunset?
- Is there enough fuel in the boat's gas tank for the return trip?

The second Jasper adventure, "Rescue at Boone's Meadow", re-introduces the distance-time-rate considerations from the first Jasper adventure, by featuring a wounded eagle and an ultralight airplane that can be used to rescue the eagle. Students are asked to find the fastest way to rescue the eagle and state how long it would take. They also have to consider issues of fuel and its effects on the payload limits of the ultralight.

The major design principles underlying the Jasper series and their functions are the following:

- *Video-based presentation format.* According to the CTGV, there are reasons to situate instruction in a video-based format, mainly because the medium allows a richer, more realistic, more dynamic presentation of information than textual material. At the same time the video-based format has some advances over real-life contexts, simply because these latter methods are not always practical, efficient, and well-structured and difficult to organize in a school situation.
- *Narrative format.* The presentation of the problem in the form of a story helps pupils to create a meaningful context.
- *Generative structure.* By having the students themselves generate the resolution of the story, their active involvement in the learning process is stimulated.
- *Embedded data design.* By having the information that will be relevant to the solution embedded in the story, students are enabled to take part in problem identification, problem formulation and pattern recognition activities that traditional word problem solving does not allow.
- *Problem complexity.* By posing very complex mathematical problems – sometimes comprising of more than 15 interrelated steps or subproblems – students are provided with the opportunity to engage in a kind of sustained and applied mathematical thinking that traditional curricula rarely offer.

Initial studies about the implementation and effect of the Jasper series
A baseline study (CTGV, 1997; Van Haneghan et al., 1992) revealed that even above-average sixth-graders were very poor in their approach to complex application problems of the kind used in the Jasper series without instruction and mediation. The data showed that students had the mathematical knowledge necessary to solve the Jasper problems, but had difficulty identifying and defining subproblems on their own. According to the authors, this was not too surprising, as traditional mathematics instruction tends not to prepare students for complex problem formulation and modeling activities.

Following the baseline study that did not involve any instruction in Jasper, a number of teaching experiments were executed that assessed the learning and transfer effects of different approaches to instruction with the Jasper series. One study involved pupils from a high-achieving fifth-grade class. On the first day of the experiment, the Jasper video was shown to all pupils and then they were pre-tested. After pre-testing, pupils were assigned either to an experimental or a control group and both groups received three additional one-hour teaching sessions. During these sessions the experimental group engaged in problem analysis, problem detection, and solution planning to check Jasper's trip-planning decision, thereby intensively relying on videodisc/computer technology controlled by the instructor. In the control group, traditional teaching methods were used to instruct students in solving traditional one-step and two-step word problems in unrelated contexts that involved the same mathematical concepts as the overall Jasper adventure. Following instruction, pupils received several post-tests to assess learning and transfer. One test was designed to capture students' mastery of the solution to the problem they had seen on the video. Not surprisingly, students in the Jasper group scored much better on this mastery test. In the second test, the pupils were administered a video-based problem the solution of which was isomorphic to the initial Jasper adventure, but embedded in a different story. Students watched this transfer video and then solved the problem while talking aloud. The analysis, involving not only the students' final answer but also how they had considered the different subgoals, indicated that Jasper students achieved much higher transfer scores than the word-problem students. Finally, all students received a post-test on one-step and two-step problems similar to those practiced in the word-problem group, and no differences were found on these problems between the groups.

Some years later the implementation and effectiveness of working with the Jasper series was investigated in a field trial involving teachers and students from nine different states. Paper-and-pencil instruments were developed for assessing three aspects of their implementation: classroom instructional activities, student outcomes, and teachers' reactions to the implementation. Data coming from a diversity of sources (teacher self-reports, on-site observations, and so on) revealed that there was wide variation in how teachers implemented the instructional model. Four test instruments were developed and administered in the Jasper classes and in appropriate comparison classes to examine the effects of the Jasper program. Both groups improved at the same rate on the Basic Math Concepts Test. On a Word Problem Test, which could be considered as a near-transfer test, the performance of the students from the Jasper classes was superior to that of the control students. On the Planning Test, which assessed higher-level planning and subgoal comprehension, students from the Jasper classrooms scored much higher at the end of the year on both aspects than students from the comparison classrooms. Finally, students' attitudes toward mathematics, assessed with a 35-item questionnaire, revealed improved attitudes of the Jasper students as compared to the comparison students on four of the five "Attitudes Towards Math" subscales. Jasper students showed less anxiety towards mathematics, were more likely to see mathematics as relevant for everyday life, more likely to see it as useful, and more likely to appreciate complex challenges. Interestingly, qualitative data from teachers commenting on the Jasper implementation project indicated that the Jasper adventures had, according to almost every teacher, a very positive effect not only on high-achieving students but also on students who previously had performed poorly in mathematics.

Continued research and development work around the Jasper series
While the overall picture emerging from all these data was that the researchers and the teachers were seeing many positive benefits from Jasper, these benefits were not as great as the researchers of the CTGV felt they could be. They were especially concerned with the problems that: (1) the flexibility with which the students could transfer their thinking to new situations was still too restricted, and (2) that they had learned a particular strategy for solving a particular Jasper adventure without acquiring deep understanding of important concepts related to that strategy. Accordingly, another line of research was set up which focused in the issue of helping students to deepen their understanding of the mathematical concepts underlying Jasper and, in the process, develop a more flexible ability to transfer to new problems. This work led the researchers to redesign the Jasper series by adding video-based analog and extension problems to each adventure. "What-if" analog problems were created by re-using the setting, characters, and objects of the videos and perturbing the values of one or more of the variables. Extension problems extended students' thinking from the adventure to other structurally similar but contextually different settings. Studies showed that opportunities to work on these analogs and extensions had strong effects on students' understanding and abilities to solve new problems that they confront.

More recently, the CTGV has further elaborated (some of) the Jasper adventures by creating multimedia-based programs (called SMART programs) which provide students and teachers with feedback about their solutions, allow them to gain access to resources to help them revise, and showcase examples of well-articulated reasoning (see also Barron, Schwartz, Vye, Moore, Petrosino, Zech, Bransford, & the Cognition and Technology Group at Vanderbilt, 1998; Vye, Schwartz, Bransford, Barron, Zech, & the Cognition and Technology Group at Vanderbilt, 1997). Each SMART program is composed of four segments:

- SMART Lab (providing summaries of, and comments on, the responses of all students participating in the learning community).
- Roving Reporter (videoclips of interviews of various students in the learning community about the problem solving they had been doing).
- Toolbox (tools for generating topic-related visual representations such as pictures, diagrams, etc. to aid problem solving).
- The Challenge (a new but related problem-solving challenge).

To look at the added benefit of the Jasper Challenge programs for student learning and attitudes, a set of nine fifth-grade classrooms from two inner-city schools participated in a 6-week study. All classes worked with the same Jasper adventure, but approximately half of the classes received the Jasper Challenge

programs and the other classes did not. In each group, students spent exactly the
same amount of instructional time, namely 18 sessions, and the teachers of all
classes (also the Jasper-only classes) received three 2-hour professional development sessions. The researchers collected data on three areas: student learning,
students' attitudes, and reactions to a the Big Challenge – a live broadcasted,
interactive event in which students had to demonstrate and apply what they had
learned during the sessions.

The results showed added value for "Jasper plus Jasper Challenges"
compared to "Jasper alone". On a transfer test involving a task that was structurally similar to the Jasper adventure but with a different cover story and different numbers, students in the Jasper Challenges group were more likely to give
correct answers and more likely to effectively justify their answers. The results
of the analyses of student attitudes revealed that the Jasper-plus students showed
greater positive changes in attitudes on two of the four scales (namely interest
and confidence). With respect to students' reactions to the Big Challenge, significant differences were found on several items of a 17-item questionnaire, indicating that students in the "Jasper plus" condition felt more prepared to tackle
the Big Challenge, thought the questions were easy, would like to participate in
other programs, and enjoyed listening to other students explaining their thinking on TV. In sum, it is clear that enriching the original Jasper environment with
sophisticated multimedia-based scaffolding tools considerably increases student
learning, transfer and attitudes when compared to classrooms that had the same
content and instruction but did not receive SMART.

Conclusion

The Jasper project is a good example of how video/computer technology can be
used to enhance mathematical modeling of complex contextualized problems
that are rich in resources and scaffolding opportunities and that offer ample
scope for social interaction. The available research evidence suggests that the
Jasper learning environment – especially in its most advanced and elaborated
form – has great potential for fostering students' mathematical modeling and
problem-solving skills, and their related beliefs and attitudes in the direction
advocated in this book.

Improving assessment

In the 1990's there was a considerable surge of activity aimed at the improvement of assessment in mathematics (e.g., Clarke, 1996; Lesh & Lamon, 1992b;
Niss, 1993a, 1993b; Romberg, 1995; Van den Heuvel-Panhuizen, 1996). In
Chapter 5 (pp. 71–73) we discussed the harmful effects that standard forms of
written assessment have on teaching and learning through the pressures that
they exert on teachers who are obligated to maximize the results achieved by
their students, and, more fundamentally, through the message that they convey
about what forms of mathematical performance are valued. As Clarke (1996,

p. 327) expressed it: "Assessment should be recognized, not as a neutral element in the mathematics curriculum, but as a powerful mechanism for the social construction of mathematical competence".

There is considerable agreement that what is formally assessed, with high-stakes consequences, drives the content and style of teaching to a considerable degree. While we have concentrated till now on the negative implications of this linkage, it is equally widely agreed that well-designed assessment has the potential to affect teaching in a beneficial way. The other side of the WYTIWYG principle ("What You Test Is What You Get", see p. 72) is that:

> Conversely, a range of high-quality tasks that assess a broader range of skills will convey messages about the nature of the desired learning activities more powerfully than any analytic description. It is hard for teachers to adopt new teaching practices, even those that offer innovative learning experiences focused on higher-level skills, if the teacher cannot see how the skills acquired will be recognized in their students. (Bell, Burkhardt, & Swan, 1992a, p. 119)

Assessment within Realistic Mathematics Education (RME)

One of the most sustained efforts to improve assessment has been carried out as an integral part of the approach to mathematics education developed at the Freudenthal Institute in Utrecht, termed Realistic Mathematics Education (RME). A comprehensive overview of the assessment aspects of this comprehensive endeavor has been provided by Van den Heuvel-Panhuizen (1996).

One principle of RME is a preference for genuine application problems, i.e., "rich, non-mathematical contexts that are open to mathematization" (Van den Heuvel-Panhuizen, 1996, p. 19). These application problems should be distinguished from traditional word problems that are "rather unappealing, dressed up problems in which the context is merely window dressing for the mathematics put there" (Van den Heuvel-Panhuizen, 1996, p. 20). By way of contrast, the following example of a test item is offered (p. 20): "Mr. Jansen lives in Utrecht. He must be at Zwolle at 9:00 Tuesday morning. Which train should he take? (Check the train schedule)". Van den Heuvel-Panhuizen (1996, pp. 20–21) comments as follows on this question:

> This problem is nearly unsolvable if one does not place oneself in the context. It is also a problem where the students need not marginalize their own experiences. At the same time, this example shows that true application problems can have more than one solution and that, in addition to written information, one can also use drawings, tables, graphs, newspaper clippings and suchlike. Characteristic of this kind of problem is the fact that one cannot learn to do them by distinguishing certain types of problems and then applying fixed solution procedures. The object here is for the student to place him or herself in the context and then make certain assumptions (such as how far

Mr. Jansen lives from the station and how important it is that he arrives at his destination on time).

The textbooks developed within the RME program emphasized application problems of this sort rather than traditional word problems, another distinguishing feature being that they were constructed out of a number of sub-problems thematically organized, rather than being fragmentary, as is typically the case with traditional word problems. Two further examples of assessment items compatible with this approach are shown in Figures 6.12 and 6.13 (Van den Heuvel-Panhuizen, 1996, p. 95 and p. 263).

As illustrated by the three examples given, important characteristics of the RME approach to application problems – including those used for assessment – that are richer than traditional word problems, include the following:

- Extensive use of visual elements – which may help to convey a situation, provide information, or even suggest a solution method – to supplement text.
- Also, the provision of various types of material (such as the train timetable in the example cited above).

Instructions to be read aloud:
"A polar bear weighs 500 kilograms. How many children together weigh as much as one polar bear?
Write your answer in the empty box.
If you like, you may use the scratch paper."

Fig. 6.12   Polar bear problem (Van den Heuvel-Panhuizen, 1996, p. 95). Copyright 1996 by the author. Reprinted with permission.

Best Buy

EVER SPORTS

discount

40%

World Sports

OUR LIST PRICES ARE
THE CHEAPEST !!!

now 25% off
our list price

a) In which of the two shops do you think the sale price of the tennis shoes is the lowest? Explain why you think so.

b) Is it also possible that the sale price of the shoes in the other shop is the lowest? Explain your answer.

Fig. 6.13   Best buys problem, with the safety-net question (Van den Heuvel-Panhuizen, 1996, p. 263). Copyright 1996 by the author. Reprinted with permission.

- All the information may not be provided. For example, in the polar bear problem (Figure 6.12) the student has to decide a reasonable value for the average weight of a child.
- There is, in general, not a single correct answer, as exemplified in the examples cited.
- By providing "scratch paper" (see Figure 6.12), indications may be elicited on the processes of tackling the problem.
- A major principle is the use of "relevant and essential" contexts (Van den Heuvel-Panhuizen, 1996, p. 91).
- Asking questions to which someone might want to know the answer.
- Asking questions that involve computations before formal techniques for those computations have been taught (and not immediately after).
- Use of "safety-net" follow-up questions that give students the chance to answer the original question in their own way (e.g., Figure 6.13).

Effects on teaching of changing assessment
As pointed out above, the influence exerted by assessment on classroom practices implies that, while it is generally considered in relation to the harmful effects produced, it has equal potential for beneficial effects, and indeed, as a

driving force for reform. In this respect, the broadening of the concept of valid-ity to "systemic validity" (Frederiksen & Collins, 1989, p. 30) is relevant:

> A systemically valid test ... is one that induces in the education sys-tem curricular and instructional changes that foster the development of the cognitive traits that the test is designed to measure ... Evidence for systemic validity would be an improvement in those traits after the test had been in place within the educational system for a period of time.

Unfortunately, it is hard to find cases where a radical reform of assessment has been carried through consistently within an educational system over a sus-tained period. For example, in the United Kingdom, despite a great deal of inno-vative and creative work on enriched forms of assessment (e.g., Bell et al., 1992a; Bell, Burkhardt, & Swan, 1992b) and a degree of implementation of wider forms of assessment, the state of affairs was described thus by Brown (1993, p. 82) (and the situation has deteriorated further since then):

> The UK now seems to be heading back to our previous position where the curriculum becomes subservient to the requirements of regular routine written examinations, which the Cockcroft Committee in 1982 identified as a major cause of low standards of motivation and achievement.

Likewise, in the United States, despite the many assessment projects of the 1990's (e.g., Resnick, Briars, & Lesgold, 1992) and the publication of the NCTM's "Assessment standards for school mathematics" (National Council of Teachers of Mathematics, 1995), the preponderance of assessment, in practice, remains unreformed (Romberg, Wilson, Khaketla, & Chavarria, 1992; Silver & Kenney, 1995).

While it is assumed by proponents of reform of assessment that such reform would have beneficial effects on teaching practice, the only systematic study of such effects of which we are aware is that reported by Clarke and Stephens (1996). They documented the effects of mandated changes in assessment in the state of Victoria on school policy and teachers' perceptions and practices. The Victorian Certificate of Education (VCE) requires students in Years 11–12 to complete (a) a multiple-choice skills test, (b) an extended answer analytic test, (c) a 10-hour "Challenging Problem", and (d) a 20-hour "Investigative Project"; all four components weighted equally in determining the final grade. Clarke and Stephens specifically investigated the "Ripple Effect" whereby they hypothesized that mandated high-stakes assessment at Years 11–12 would result in changes in mathematics teaching in Years 7–10.

The investigation proceeded through three stages:

- Stage 1: Documents relating to mathematics teaching for a sample of Victorian high schools were analyzed. This analysis revealed that to a con-siderable extent, documented policy on teaching practice and assessment

incorporated the distinctive features of the official "Mathematics Study Design" (Victorian Curriculum and Assessment Board, 1990).
- Stage 2: Through questionnaires, it was verified that high levels of support were expressed by teachers for aspects endorsed by VCE curriculum advice and assessment practice.
- Stage 3: On the basis of teacher interviews, the researchers concluded that there was a substantial commonality of meaning for those terms and practices integral to VCE mathematics.

Clarke and Stephens (1996, p. 90) claimed that "the three phases of this study substantiate the hypothesized Ripple Effect in Years 7 to 10 of changed assessment practices at Years 11 and 12". While reporting some evidence of more deep-seated motivations, they concluded that the main reason for the effect was a pragmatic reaction to the demands of the new forms of assessment. A particularly important conclusion reached was that "teachers are reluctant to embrace new assessment and instructional practices unless these are policy driven, that is, have the endorsement of inclusion in high stakes assessment" (p. 90).

Summary and discussion

This chapter began with reports on three design experiments, all of which have, to different degrees, produced positive outcomes in terms of students' performance, thinking processes, beliefs and attitudes. In retrospect, it is interesting to consider to what degree there are similarities among the three projects. First, all projects used more realistic, more challenging, and more open tasks than the traditional textbook word problems, although the Jasper project goes much further in this direction than the other two. Second, all projects apply a comparable variety of teaching methods and learner activities, including the use of expert modeling of strategic aspects of the solution of mathematical application problems, guided practice in small groups with coaching and feedback, and whole-class discussions focusing on evaluation and reflection concerning alternative mathematical models and different solution strategies. Third, all interventions aimed to create a classroom climate that is conducive to the development in pupils of appropriate beliefs about mathematics and mathematical modeling and problem solving. Finally, apart from the first study by Verschaffel and De Corte (1997b) and some of the initial studies of the CTGV, teachers were not simply provided with rich teaching and learning materials. Taking into account the crucial importance of teachers' conceptions and beliefs, at the outset of, and during, the projects attention was paid to the professional development of teachers, focussing not on procedural skills but on the understanding of the learning goals and the design principles underlying the projects.

A final crucial piece in any sustained and coherent attempt to improve mathematics teaching – besides the creation of appropriate materials, and the professional development of teachers to use those materials effectively – is the development of assessment methods that are consistent with the underlying goals. While the 1990's produced an upsurge in innovative and creative efforts to realize enriched forms of assessment, the impact on schools in general has been limited. The example of events in the state of Victoria, Australia, provides a rare example showing how changes in assessment can impact teaching on the ground.

Part 3

Word Problems from a Modeling Perspective

In the first part of this book, we described how our interest was aroused by numerous examples in the literature, from many parts of the world, of cases where students answered word problems apparently without regard for realistic considerations. The most extreme and sensational example was the finding by French researchers that children were prepared to answer nonsensical questions such as: "There are 26 sheep and 10 goats on a ship. How old is the captain?"

For the items used in the French research there is, at least from our perspective, no reasonable logical link between the text and the answers given by students, except in the minimal sense that the students often checked the size of the answer for plausibility. Other cases – which have in common that they appear to show suspension of sense-making – are different in nature in that a reasonable answer can be given, if only in the form of an approximation. It is items of this type that we have used in our research (see Table 2.2), an example being: "What will be the temperature of water in a container if you pour 1 liter of water at 80°F and 1 liter of water at 40°F into it?" Our reaction to the reported "disaster studies" led us to carry out initial investigations (Greer, 1993; Verschaffel et al., 1994) that further demonstrated the phenomenon. As reported in Chapter 2, subsequent replications by ourselves, our research colleagues and others in many countries – Switzerland, Sweden, Japan, Venezuela, Germany – yielded highly consistent results (see Table 2.5) showing that children in all these countries routinely answer many non-straightforward questions as if they are unproblematic.

As a next step, a number of studies were carried out to test the hypothesis that the students' responses could be simply explained in terms of lack of attention in some sense, and that the minimal intervention of alerting them to the possibility that the questions might have some problematic features needing interpretation would be effective. A range of results reported in Chapter 3 provided strong evidence against this hypothesis. We interpreted these highly consistent findings as indicative that the patterns of responses are deeply rooted.

By contrast, in Chapter 4, we reported a number of studies in which task presentation was modified in a more fundamental way, namely by presenting tasks in relatively authentic settings that simulated, to some degree at least, the goals, social circumstances, and realistic constraints that would influence performance on corresponding tasks carried out in real life. As we reported, these changes in the "experimental contract" markedly increased the number of responses deemed to be showing awareness of realistic considerations.

Thus, in Part 1 we concentrated on reporting observations without going in any depth into interpretations of those observations or our emerging theoretical ideas. The first stage of the research program concentrated on ascertaining studies, i.e. studies documenting the existing state of affairs (Chapters 1 and 2) and on interventions taking the form of modifications of the tasks presented (Chapters 3 and 4).

While our initial reaction to the first findings was one of amazement at the apparent irrationality of the children's responses, we progressively realized that this was a naive interpretation as we continued to study existing literature, and discussed the findings amongst our research associates and others in the course of conference presentations. A key insight was that behavior that, at first sight, appears irrational, can be seen as rational if considered against the background of schooling in general, and mathematics classrooms in particular. Thus, according to Reusser and Stebler (1997a, p. 325):

> As a result of schooling, students' behavior is pragmatically function-
> al if they take into account any information they can draw from both
> problem texts and contexts. That is, their mathematical sense-making
> is functional if they actively and continuously construct a mental rep-
> resentation not only of the specific task (problem model ...) but also of
> the socio-contextual situation which they are in (... social context
> model).

In line with this growing sense that the crucial factor was the situatedness of word problem solving within classroom culture, Part 2 of the book reflects the broadening of both our theoretical perspective and our experimental program, through focussing on the educational environment and through a shift to teaching experiments and analysis of exemplary alternative instructional environments.

Chapter 5 began with a detailed analysis of the way in which word problems are currently taught in typical mathematical classrooms, drawing on a substantial body of observations, experimental evidence, and theoretical reflection. Among the strongly influential shaping factors that can be identified are textbooks and assessment. However, the most important factor, arguably, is the role of the teacher. As a first step in following this surely essential line of future research, a study of pre-service teachers' performance on the same word problems given to students, and of their views about these word problems, was reported. The central finding was that these future teachers showed, to an alarming extent, the same tendency as students to disregard aspects of reality when responding to problematic items. Moreover, their ratings of different types of response to such items indicated that, to a considerable degree, they considered it reasonable for students to respond to such items in this way.

In Chapter 6, studies aimed at radically changing students' perceptions of word problem solving were reported. The first two were carried out by the research group at Leuven and involved carefully planned teaching experiments – the first carried out by a researcher, and the second by teachers in a number of schools. These interventions, informed by reflection on the findings of previous research, addressed all components of successful problem solving in relation to mathematical application problems. They promoted problem solving at both strategic and tactical levels and presented carefully tailored examples in order to raise awareness of the need for realistic modeling of situations. A variety of teaching methods and learner activities were used, including expert modeling of strategic approaches to the solution of application problems, small-group work with coaching and feedback, and whole-class discussions reflecting on alternative models and solution strategies. All of these aspects aimed at establishing a very different set of "socio-mathematical norms" (Cobb, 1996) than prevail in classrooms. In recognition of the vital role of teachers, in the second intervention considerable effort was put into aligning the teachers with the learning goals and design principles of the study.

Next, a more sustained program of research carried out by the Cognition and Technology Group at Vanderbilt was reviewed. In terms of the underlying philosophy, we consider this program to share important principles with our work, most centrally the aim of helping students achieve an ability to model complex situations. The CTGV program is distinguished by innovative use of video/computer technology that allows the presentation of dramatically presented, narratively coherent, informationally rich scenarios. Finally, in Chapter 6, bearing in mind the accepted view that teaching, to a considerable degree, is driven by assessment methods, illustrative examples of approaches to assessment compatible with our views on the importance of realistic modeling were presented, together with a summary of a study carried out in Australia demonstrating that the effects of assessment on teaching, usually discussed in the context of negative influence, may also be positive.

Thus, the findings of our research program, together with cognate research, reported in the first two parts of the book, present a consistent picture of how students typically solve word problems without regard to realistic considerations and document examples of the impact of alternative instructional environments.

Throughout the course of the experimental studies, of course, our theoretical framework was developing. Moreover, it is essential to acknowledge that both our evolving theoretical ideas and our experimental work are shaped by a view of the nature of mathematics and value judgments about the aims of mathematics education. In Part 3, accordingly, we step back from the reporting of research findings to undertake a wider discussion of philosophical and theoretical issues, further analysis of the features of the educational system that lie at the root of what we consider to be outcomes seriously detrimental to many students' understanding and conception of mathematics, and suggestions for a radical reconceptualization of the role of word problems within the curriculum.

Among the fundamental questions underlying our research is the complex relationship between mathematics and reality. In particular, we focus on the use of mathematics to model aspects of the real world. From this perspective, as discussed in Chapter 7, word problems can be construed as (more or less simple) exercises in mathematical modeling, as opposed to routine applications of standard arithmetical procedures, a view advanced from the beginnings of this research and, indeed, earlier (Greer, 1987, 1992) and central to the design and implementation of the teaching studies carried out.

In Chapter 8, we turn to analysis of word problems – that "peculiar cultural device" as Lave (1992, p. 75) described them. They are considered in a first section from the historical perspective, reviewing how they have been used in mathematics education with remarkable stylistic homogeneity across time and cultures. An important reason for considering the historical perspective is that it may throw some light on the fundamental question "What are word problems *for*?" We suggest that there are two, highly contrasting, main motivations for their use. On the one hand, they are regarded as providing examples of how mathematics can be applied to real-world problems, including those within the everyday present and future experience of students (e.g. Burkhardt, 1981). On the other hand, they may be presented as (mostly implicitly) artificial exercises providing frameworks within which mathematical structures may be explored. Toom (1999, p. 36) characterizes these two uses of word problems as "applications" and "mental manipulatives". While fully acknowledging the importance of the latter role for word problems, it is with the former that we are primarily concerned – more specifically, with word problems that ostensively relate to the real world but, in one way or another, fail to do so realistically.

In the next section, we turn to word problems as linguistic entities (which by definition they are), subject to a variety of analyses as "a very special type of text" (Nesher, 1980, p. 41). One major reason for unrealistic responding to word problems is simplistic mapping of linguistic frames onto mathematical structures, a particularly clear example being in relation to proportional reasoning. Next, a variety of theoretical perspectives and constructs from linguistic theory as applied to the analysis of word problems (e.g., Gerofsky, 1996; Pimm, 1995) are considered. These analyses lead to a consideration of the relationship between text and the situations it purportedly describes, as a central aspect of the modeling process discussed in Chapter 7, and of the distinction between word problems as applications and word problems as mental manipulatives. We then return to the nature of the immediate school/class context in which students encounter word problems, as discussed at length in Chapter 5 as the underlying reason for observed student behavior, and in Chapter 6 in the context of examples of radically different instructional environments. In this respect, Lave's (1992) characterization of mathematics in school as a form of situated cognition is useful both as a means of explaining documented findings and in recognizing that change is possible.

In the last section of Chapter 8, we acknowledge the lack in our research to date of attention to a number of important characteristics of students that interact with their views of, and approaches to, word problems – in particular, gender and class.

In Chapter 9, we end the book with radical suggestions for reconceptualizing the role that word problems might play in mathematics education. While making these recommendations we are mindful of the many complexities and difficulties that would be involved, and the massive investment of human resource that would be required, yet the starting-point is our contention that our work, as well as much of recent research in the field of mathematics education, has clearly demonstrated the need for change. Our key proposals are to adopt a modeling perspective, to increase the dialectical richness of interchanges between students and teachers and among students, and to make more explicit the rules of the word problem game. While word problems are the focus of our work, in many ways the questions they raise and the answers that may be suggested stand as representative of the diagnosed ills and proposed remedies for mathematics education in general.

7

Mathematical Modeling of Aspects of Reality

A man was drinking with his friends. He told them that he had been to see his doctor, who had told him to cut down on alcohol consumption. "She told me I could drink two pints of beer every night", he reported. As he started his fourth pint, his friends reminded him of the doctor's advice. "That's all right", he replied, "I wanted a second opinion, so I went to another doctor. He also said I could have two pints every night". (Joke of unknown origin)

In this chapter, we consider a fundamental question relating to mathematical application problems in general, and word problems in particular, namely the relationship between aspects of reality and mathematical structures. We argue that a simplistic view of this relationship is a major factor contributing to the phenomenon described and probed in the first two sections of this book, namely a tendency to assume that all application problems can be appropriately responded to by simple mathematical formulations, while ignoring complications inherent in the real-life situations described.

We propose a characterization of mathematics as having two faces, one being its role in describing aspects of the world, and the other the construction of abstract structures; modeling forms the link between these two aspects. Within such a framework, we illustrate how word problems can be reconceptualized as exercises in mathematical modeling.

Relationship between arithmetic statements and aspects of reality

Soon after beginning formal schooling, children become familiar with statements of elementary arithmetic, such as $7 + 3 = 10$, $3 \times 4 = 12$, with manipulatives and visual representations of such statements, and with corresponding word problems such as:

- John has 7 apples and Mary has 3. How many do they have altogether?
- How many cookies are there in 3 boxes with 4 cookies in each box?

The apparent simplicity of such statements contrasts with a very substantial body of research showing the psychological complexity of the development of understanding of such computations and their application to highly varied situations (e.g., Carpenter, Moser, & Romberg, 1982; Fuson, 1992; Greer, 1992; Harel & Confrey, 1994; Hiebert & Behr, 1989; Verschaffel & De Corte, 1993, 1996).

The meaning of statements in arithmetic has been subjected to intense philosophical analysis through the ages. A range of opinions on the subject is discussed by Hersh (1997), from the declaration by St. Augustine that:

> Seven and three are ten, not only now but always; nor was there a time when seven and three were not ten, nor will there ever be a time when seven and three will not be ten. I say, therefore, that this incorruptible truth of number is common to me and to any reasoning person whatsoever ... (cited by Hersh, 1997, pp. 103–104)

to Wittgenstein's contention that $3 + 5 = 8$ only because "That's how we do it" (Hersh, 1997, p. 202).

St. Augustine's view is the predominant one. "Everyone knows" that $2 + 3 = 5$ because if you put 2 apples on a table and then another 3 apples and count all the apples, you will find there are 5 and that $3 \times 2 = 6$ because there are 6 objects in 3 groups of 2 objects. It is standard to introduce children to arithmetical computation by having them perform operations such as combining, separating, comparing groups of objects such as blocks. The experience of doing things with relatively small collections of discrete objects that afford counting becomes internalized as a generalized "image" (Johnson, 1987) for natural numbers and operations on them (Toom, 1999).

From this common-sense perspective, the anecdote with which this chapter opens is only a joke – or is it? Wittgenstein (1956, p. 14e) used the following example:

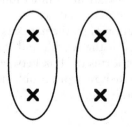

You only need to look at the figure to see that 2 + 2 = 4.
– Then I only need to look at the figure

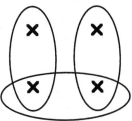

to see that 2 + 2 + 2 = 4.

Indeed, the specific statement 2 + 2 = 4 has received very extensive discussion. Bloor (1994) discusses it from the perspective of sociology of knowledge, Ascher and Ascher (1997, pp. 29–30) from the perspective of ethnomathematics, Lakoff and Nunez (1997) in terms of metaphorical cognition, while Restivo (1993) on the same topic cites authors as diverse as Kline, Thomas Hardy, Karl Mannheim, George Orwell, Dostoevsky, Oswald Spengler, Bertrand Russell, and Quine.

The example from Wittgenstein just discussed draws our attention to the fact that addition is only appropriate for the combination of sets of objects when those sets do not overlap. Consider the first pair of items used by Verschaffel et al. (1994) and discussed in Chapter 2:

> S1 Pete organizes a birthday party for his tenth birthday. He invited 8 boy friends and 4 girl friends. How many friends did Pete invite for his birthday party?

> P1 Carl has 5 friends and Georges has 6 friends. Carl and Georges decide to give a party together. They invite all their friends. All friends are present. How many friends are there at the party?

Here the disjunction of the sets is implied in the statement of the first problem, but in the second it is possible, indeed likely, that Carl and Georges will have some friends in common. Accordingly the result is indeterminate, ranging from 6 to 11 (not counting Carl and Georges). In the seven studies reviewed in Chapter 2, the number of children who showed any awareness of this indeterminacy ranged from 3% to 23% (see Tables 2.4 and 2.5).

A more complex, but related, example is P6:

> Bruce and Alice go to the same school. Bruce lives at a distance of 17 kilometers from the school and Alice at 8 kilometers. How far do Bruce and Alice live from each other?

Again, the answer is indeterminate – any distance between 9 and 25 kilometers is possible. In the seven studies (Tables 2.4 and 2.5) only 1% to 5% of children indicated awareness of this indeterminacy.

Thus, initial learning of addition (and subtraction) is aided by manipulations and representations of sets of small, discrete objects and operations such as putting together discrete sets, comparing two sets, and taking away a subset. However, it is important at the same time to be able to discriminate other situations in which addition (or substraction), at first sight, might appear to be appropriate, but is not.

A case in point is the following item:

> What will be the temperature of water in a container if you pour 1 jug of water at 80°F and 1 jug of water at 40°F into it? (Nesher, 1980, p. 46)

This is an example where "putting together" does not match the arithmetical operation of addition. The same item was used (as P3, Table 2.2) by Verschaffel et al. (1994), with only 17% realistic reactions, and in four replications the percentage of RRs ranged from 9% to 21%. Nesher (1980, p. 46) suggests the explanation is that the children are following an explicit or implicit rule that "when you put things together, you add the numbers".

The origins of children's understanding of arithmetic operations lie in sensori-motor experience of physical, represented, or imagined manipulations of objects – such as putting sets together, removing some objects from a set, creating multiple equal groups, sharing equally, and so on. The result of this activity is the emergence of "primitive models" for addition, subtraction, multiplication, and division whereby, according to Fischbein, Deri, Nello, and Marino (1985, p. 4): "Each fundamental operation of arithmetic generally remains linked to an implicit, unconscious, and primitive intuitive model" (see also Johnson, 1987).

Davis and Hersh (1981, p. 71) pose the following question, which could be interpreted as a simple application of multiplication:

> One can of tuna fish costs $1.05. How much do two cans of tuna cost?

Their grocer, it turned out, sold two cans for $2.00. Making the price of something directly proportional to the quantity is reasonable and used in many instances, but by no means all; it is a law neither of nature nor of humankind. Ascher and Ascher (1997, p. 29) relate the anecdote of an African sheepherder and a European. The herder agreed to accept two sticks of tobacco for one sheep but was reluctant to accept four sticks of tobacco for two sheep. From a naive perspective, this may be interpreted as evidence that the herder did not understand that 2 + 2 (or 2 × 2) = 4. In fact, as Ascher and Ascher suggested, "the problem is not that the shepherd doesn't understand arithmetic, it is rather that the scientist/trader doesn't understand sheep", the point being that it cannot be assumed that one sheep is equally as valuable as another.

Quite apart from such examples, where the application of direct proportionality is a matter of choice or convention, there are many cases where modeling by direct proportionality is totally inappropriate, or at best offers a very rough approximation, such as these two items introduced and discussed in Chapter 2:

P5 John's best time to run 100 meters is 17 seconds. How long will it take him to run 1 kilometer?

P10 This flask is being filled from a tap at a constant rate. If the depth of the water is 4 cm after 10 seconds, how deep will it be after 30 seconds?

In the several studies in which these items were used (see Table 2.5), the number of children showing awareness that direct proportionality will give an approximate answer at best ranged from 0% to 7%, and from 0% to 4%, respectively. These results illustrate how strong is the "illusion of proportionality" (De Bock, Verschaffel, & Janssens, 1998; for other examples, see Puchalska & Semadeni, 1987, as discussed in Chapter 1 (p. 10)).

Another way to put the "obviousness" of $2 + 3 = 5$ and $3 \times 2 = 6$ into perspective is to compare them with other arithmetical statements like $3 - 5 = -2$, $-3 \times -2 = 6$. Young children will often say that "you cannot take 5 from 3" (so will teachers when teaching multicolumn subtraction) and, in concrete terms, they are, of course, right. If asked to take 5 from 3, 0 is sometimes the response; if you have three apples on a table, and are asked to take five away, that's about the best you can do! Negative numbers do not arise naturally within a context of operations such as combining and separating collections of countable objects. As far as additive operations go, they do provide a way of mathematizing other situations, such as debit and credit (if you have 3 coins and owe 5, you can issue a promissory note for 2) or height above and below sea-level, and can be represented in various ways, such as displacements along a number line (e.g., Vergnaud, 1982). Nevertheless, the historical record shows the reluctance of even strong mathematicians (e.g., Augustus De Morgan) well into the 19th century to accept negative numbers. When it comes to the multiplication of negative numbers, the situation is even more complex. While certain situations can be described to make it plausible why -3×-2 is defined as $+6$, many (e.g., Fischbein, 1987) argue that the reasons for adopting that definition are ultimately based on criteria that are formal (maintaining consistency of properties) and pragmatic (usefulness).

Hersh (1997) has suggested a plausible way to resolve the philosophical issue of the "meaning" of a statement like $2 + 2 = 4$ – by arguing that it has not one meaning, but two. On the one hand, it is a statement about mathematical objects – objects that, Hersh contends, are neither physical nor mental, but social constructions. On the other hand, $2 + 2 = 4$ can be taken as a description of an aspect of physics, that "two discrete, reasonably permanent, noninteracting objects collected with two others make four such objects" (p. 15). Hersh concludes (p. 16):

> So "two" and "four" have double meanings: as Counting Numbers or as pure numbers. The formula $2 + 2 = 4$ has a double meaning. It's about counting, about how discrete, reasonably permanent, non-interacting objects behave. And it's a theorem in pure arithmetic ...

Relationship between mathematics and reality: The role of modeling

Hersh's proposed solution to the philosophical question of the nature of arithmetical (more generally mathematical) statements is consistent with a view of mathematics as having a dual nature (De Corte et al., 1996, p. 500):

> On the one hand, mathematics is rooted in the perception and description of the ordering of events in time and the arrangement of objects in space, and so on ("common sense – only better organized", as Freudenthal (1991, p. 9) put it), and in the solution of practical problems. On the other hand, out of this activity emerge symbolically represented structures that can become objects of reflection and elaboration, independent of their real-world roots.

The interaction between these two aspects of mathematics is typically complex. Mathematics developed as a purely abstract activity has been found in many cases, often after a considerable period of time, to be usefully applicable. The history of the mathematical theory of symmetry offers a particularly rich example (Stewart & Golubitsky, 1992; Weyl, 1969). With its origins in the perception of biological and other forms of natural symmetry, and in aesthetics, the mathematical theory of symmetry led into group theory and back into applications such as crystallography.

The link between the two faces of mathematics is modeling; aspects of reality can be modeled by mathematical structures. (Conversely, in another sense of "model", naturally occurring regularities or designed embodiments may be said to furnish models for mathematical structures (Gravemeijer, 1997), but that is not our concern here).

Schematically, the steps in the modeling process, as shown schematically in Figure 0.1 (p. xii), are as follows (cf. Burkhardt, 1981):

- The starting point is some aspect of reality considered potentially capable of mathematization. In authentic tasks, this aspect arises from an actual situation under analysis. More usually, in school, a description of some situation is provided – in the case of word problems, a (frequently very simplified) verbal description. By way of example, consider the following (that could arise as an authentic question for analysis, or be posed verbally): "How fast should cars drive through a single-lane tunnel?"

- The first stage in modeling is the construction of a mental model, possibly mediated by external representations, reflecting the key variables in the situation, the temporal and causal relations between them, and so on. Following Kintsch and Greeno (1985) and Reusser (1990), we term this a "situation model". In the case of the example, there are numerous factors to be considered, including length of cars and distance between them (it would be natural to draw a diagram to represent these), speed, stopping distances for drivers, and so on.
- At the next stage, mathematical equations are set up that reflect the situation model. For the example, equations can be set up relating speed, distance between cars and car lengths, safe stopping distances, and so on. Often, as in this case, simplifying assumptions may be appropriate, such as taking a fixed length for the "typical" car, although in reality cars differ in length.
- Mathematical derivations from the equations are worked out. For the example, one output could be a graph showing how an assumed steady speed should be related to distance between cars, taking into account some model of safe stopping distance.
- The mathematical derivations are interpreted. In the example, this may mean recommending a particular steady or maximum speed for driving through the tunnel.
- The proposed solution is evaluated relative to the situation model. Does it appear to make sense? If not, the situation model may be reconsidered, leading to a revised mathematical model and another cycle of the process.

In recent years, there has been a pronounced shift in many countries towards the teaching of mathematics through applications (de Lange, 1993, 1996, 1998; Keitel, 1993) as reflected, for example, in the *International Conference on the Teaching of Mathematical Modeling and Applications* held biennially, and the associated publications (e.g., Houston, Blum, Huntley, & Neill, 1997). For example, Keitel (1993, p. 20) wrote as follows:

> About ten years ago, mathematics educators propagating the teaching of mathematics by applications represented a unique group. Emphasizing the value and necessity of applications in mathematics education was judged as theoretical and practical action against the dominance of pure mathematics in schools raised by the New Math movement. Now, in 1991, it seems that such rebels have become rather a fashionable group, trendsetters in successfully propagating modeling and application in classroom practice ...

De Lange (1993, 1996, 1998) has analyzed the arguments for, and problems with, this shift in balance in school mathematics. In line with the approach of

the Freudenthal Institute, he emphasizes that situations open to mathematization can be used, not just for the application of already developed mathematics,
but as contexts for the development of mathematical concepts (cf. Gravemeijer,
1997; Van den Heuvel-Panhuizen, 1996). Thus, the relationship between situations and models is complex and interactional.

Word problems from the modeling perspective

Traditionally, and in most cases up to the present day, word problems have been
presented as mapping directly and unproblematically onto simple arithmetical
statements. For example, Säljö (1991) discussed the following problem, which
appeared in the Treviso arithmetic of 1478 (Swetz, 1987, p. 163): "If 17 men
build 2 houses in 9 days, how many days will it take 20 men to build 5 houses?"
    The solution is derived by twice using "the rule of 3" (to find one unknown
quantity out of four proportionately related) and is given as 19 days and 3
hours. Säljö comments (p. 262):

> It is clearly presumed for the task to function as a suitable exercise
> that the outcome in productivity from one man, whether working in
> a group of 17 or in a group of 20, is not affected. Similarly, it is tac-
> itly assumed that whether one is building 2 or 5 houses it makes no
> difference in terms of efficiency. The meaning has to be established
> within the context of a paper version of the world and important
> aspects of what would be the appropriate way of specifying mean-
> ing in a different setting have to be bracketed when dealing with
> these statements as exercises in arithmetic.

Moreover, the answer given implies that "day" is interpreted as 24 hours, leading to this comment (p. 263):

> Arguments rooted in an external reality, in which people do not
> work for 24 h[ours] a day can – and have to – be temporarily disre-
> garded; if they are not the problem becomes difficult to handle.

A contrasting analysis of a rather similar problem was provided by Lewis
Carroll as long ago as 1880 (Fisher, 1975). This is the problem: "If 6 cats kill 6
rats in 6 minutes, how many will be needed to kill 100 rats in 50 minutes?" and
this is what Carroll said about it:

> This is a good example of a phenomenon that often occurs in work-
> ing problems in double proportion; the answer looks all right at first,
> but, when we come to test it, we find that, owing to peculiar cir-
> cumstances in the case, the solution is either impossible or else indef-
> inite, and needing further data. The "peculiar circumstance" here is
> that fractional cats or rats are excluded from consideration, and in
> consequence of this the solution is, as we shall see, indefinite.

Having shown that the "solution" obtained by assuming direct proportionality is 12, he continued:

> But when we come to trace the history of this sanguinary scene through all its horrid details, we find that at the end of 48 minutes 96 rats are dead, and there remain 4 live rats and 2 minutes to kill them in: the question is, can this be done?
>
> Now there are at least four different ways in which the original feat, of 6 cats killing 6 rats in 6 minutes, may be achieved ...
>
> A. All 6 cats are needed to kill a rat; and this they do in one minute, the other rats standing meekly by, waiting for their turn.
>
> B. 3 cats are needed to kill a rat, and they do it in 2 minutes.
>
> C. 2 cats are needed, and do it in 3 minutes.
>
> D. Each cat kills a rat all by itself, and takes 6 minutes to do it.
>
> In cases A and B it is clear that the 12 cats (who are assumed to come quite fresh from their 48 minutes of slaughter) can finish the affair in the required time; but, in case C, it can only be done by supposing that 2 cats could kill two-thirds of a rat in 2 minutes; and in case D, by supposing that a cat could kill one-third of a rat in 2 minutes. Neither supposition is warranted by the data; nor could the fractional rats (even if endowed with equal vitality) be fairly assigned to the different cats. For my part, if I were a cat in case D, and did not find my claws in good working order, I should certainly prefer to have my one-third-rat cut off from the tail end.

Carroll showed that, while method A or B leads to the answer 12, C leads to the answer 14 and D to the answer 13.

The wit of this analysis should not disguise the important insights it implies (just as Lewis Carroll's amusing use of language conceals an underlying linguistic theory of considerable subtlety). What he has done is to replace an unreflective application of a computational procedure by an analysis that:

- Makes specific the assumptions underlying the subsequent arithmetic – in fact, a number of alternative sets of assumptions.
- Recognizes that these assumptions cannot be derived from the data stated in the problem alone but are based on what one believes about the situation described, and accordingly ...
- Takes into account (albeit in whimsical fashion) the biological and physical reality of cats and rats.

Between the direct application of precise proportionality, with no hint of any problematic aspect, of the Treviso arithmetic example (and other similar examples in the same source) and Carroll's discussion of the cats and rats there

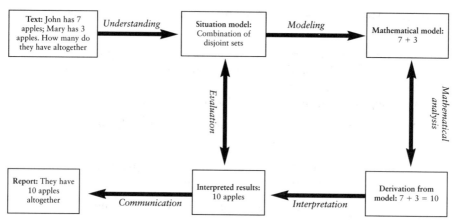

Fig. 7.1   Construing a simple word problem as an exercise in modeling.

is a fundamental change of perspective. Säljö's analysis of the house-building problem parallels that of Carroll in examining assumptions and talking about the reality of people building houses (who do not work 24 hours a day, for example).

Even the simplest word problem can be viewed as a modeling exercise, in which the phenomenon under investigation is represented by text. Consider the first example cited in this chapter: "John has 7 apples and Mary has 3. How many do they have altogether?". The relationship in this example between the situation as described and the corresponding arithmetical expression is so obvious that it takes a conscious effort to construe the process of solving the problem as a modeling exercise (Figure 7.1). We may say that this is an example where the simple arithmetic operation *does* provide an appropriate model, as is the case with the items labeled as S-problems in our research (Table 2.2).

From this perspective the basic arithmetical operations can be viewed as providing possible models for situations. The variety of such situations has been thoroughly analyzed in the literature. For example, a distinction has been made between three major "semantic schemas" for addition and subtraction, namely Combine, Change, and Compare problems (Riley et al., 1983). This classification has been expanded. Indeed, according to Davis and Hersh (1981, p. 74) "there can be no comprehensive systematization of all the situations in which it is appropriate to add". Likewise, multiplication and division are applicable to many situations, including (to use the classification suggested by Greer (1992) equal groups, equal measures, rate, measure conversion, multiplicative comparison, part/whole, multiplicative change, Cartesian product, rectangular area, product of measures. The problem, as strikingly demonstrated by our findings, is that children's responses do not show any discrimination between such problems and those where the simple arithmetic operation *does not* provide an appropriate model, of which the P-problems (Table 2.2) are examples.

One approach to counter this tendency to treat all word problems as S-problems, so to speak, is to use word problems and situations where the need for modeling becomes clear. The following example of such an activity (Lesh, personal communication) is based on a situation that actually occurred at the monthly school board meeting in a small (American) town. The issue was how to find a way to rank students' art work. The art department had, for two years, been using projects, with each student having the chance to complete up to 30 projects during the year. On the basis of this work, marks were awarded for "quantity", based on the number of projects completed, and "quality", based on the number of tools and techniques used. The question at issue was how to combine these two scores in order to produce a single ranking. The account continues:

> The debate became heated when participants at the meeting couldn't agree how the two scores should be combined. Five different procedures were suggested, and more seemed possible. Arguments arose because, when the two scores are combined in different ways, different students ranked high and low.
>
> One administrator suggested that the two scores should simply be added [QUALITY + QUANTITY] because "That's what you almost always do with two scores!".
>
> But the tennis coach suggested that it might make more sense to subtract to find differences between the two scores [QUALITY − QUANTITY] because "That's what you do, in fields like tennis, when you want to factor out one of the scores".
>
> An ex-math teacher on the school board pointed out that the two scores could also be multiplied to produce a score that means "total quality" [QUALITY × QUANTITY].
>
> But another teacher suggested, that the two scores could also be divided to produce a score of quality-per-unit-quantity [QUALITY ÷ QUANTITY].
>
> Finally, a physics teacher argued that the original two scores are two completely dimensions. So, the Pythagorean Theorem should be used to calculate the length of the vector sum [$\sqrt{(\text{QUALITY}^2 + \text{QUANTITY}^2)}$].
>
> To close the debate, the superintendent formed a committee to investigate the suggestions that had been made. Its goal is to make a recommendation about what policy should be adopted at the next meeting.

This is a good example of what Lesh calls a "model-eliciting activity" (for several others, see Lesh & Lamon, 1992a). The scenario can be presented to students with a request to formulate a recommendation to the committee as to what method of combining marks to use, with supporting arguments. A number

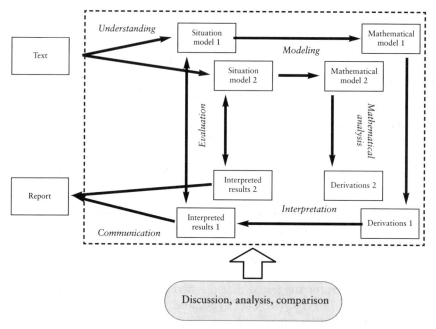

Fig. 7.2   Comparing competing models.

of important points are illustrated by the example. In particular, there is no obvious "right answer". Rather, there are a number of candidate models, yielding different results with important implications, as is typically the case in modeling of complex real-world social phenomena. In these circumstances, it becomes necessary to evaluate the relative merits of different models, on the basis of discussion, as schematically shown in Figure 7.2 (in which, for simplicity, only two alternative models are indicated).

Another example of a modeling exercise suggested by Kilpatrick (1987) was discussed in Chapter 1 (pp. 11–12 and Figure 1.2). Initial models for the amount of clothesline needed for the two configurations of racks might be simply seven lines each 6 feet long and three concentric squares of sides 2, 4, and 6 feet, respectively, yielding answers of 42 feet and 48 feet, respectively. As Kilpatrick points out, both models are unrealistic in that they fail to take into account the amount of line needed to tie the line to the supports. A second pair of models might assume an extra 3 inches for each knot, leading to answers of 45½ feet and 54 feet, respectively. A further iteration of the modeling process might involve using one continuous line knotted only at each end.

The rope item used in our research is a much simplified problem inspired by this example. In seven studies reported in Chapter 2, the number of children whose responses showed any indication of taking account of the amount of rope needed to tie the knots ranged from 0% to 8% (Table 2.5). In relation to the modeling process, Kilpatrick (1987, p.131) commented as follows:

People who want to apply their mathematical knowledge to solve a real-life problem find that the formulation of the problem in terms that will permit a mathematical solution is typically a more challenging and creative endeavor than obtaining the solution itself ... Much work remains to be done in studying the conditions under which students can come to accept a problematic situation as presenting a legitimate challenge to their mathematical abilities rather than as simply another artificial school task.

By way of illustration, let us consider a variety of ways in which the runner problem and the flask problem might be answered more adequately than by direct proportionality (which was the response on both items of over 95% of students in the studies reported in Chapter 2 (Table 2.5)).

Possible approaches to the runner problem would be:

- To give the answer 170 seconds as an approximation, with a comment that the time would certainly be greater than this because a runner cannot keep up for a long distance the maximum pace possible for a shorter distance.
- The answer might be estimated, say as 190 seconds.
- One way to produce a more accurate approximation would be to carry out an experiment with several children, in each case measuring the ratio of their times for 1 kilometer versus 100 meters. This ratio could then be applied to John's time (though the assumption underlying this solution should be noted).
- Another way would be to look up records (Olympic winning times, say) and derive the ratio from these data.

For the flask problem:

- The simplest answer recognizing that the water will not rise at constant height in the flask would be to give 12 cm. as an approximation, again with a comment that it would certainly be higher.
- The answer might be estimated, say as 15 cm.
- The answer might be worked out through experimentation with similar flasks.
- A geometric answer is possible, given the dimensions of the flask. Working out the volume of a flask of this shape, with circular cross-section (given the formula for the volume of a cone) is a problem similar to that discussed by Polya (1957) of finding the volume of a frustrum of a pyramid.

Summary

In this chapter, we have considered a fundamental epistemological problem, namely how the abstract structures of mathematics relate to aspects of phenomena in the real world. The link between the two faces is modeling.

We argued that children at an early stage in formal education, laying the foundations for understanding of arithmetical operations, encounter examples in which the mathematics relates in a perceptually and conceptually straight-forward way to the situations being explored or manipulated. As formal edu-cation proceeds, the assumption of simple relationships is reinforced through the stereotyped nature of word problems and the routine answers expected, with the results that have been documented in previous chapters.

In traditional word problems, going back many centuries, mathematical pro-cedures such as the Rule of Three tended to be applied to situations without consideration as to the realistic constraints, as in the example discussed by Säljö (1991) from the Treviso arithmetic of 1478. It takes a radical conceptual shift to move from the uncritical application of a simple and neat mathematical for-mula to the modeling perspective that takes into account the reality of the situ-ation being described. By way of illustration, a humorous, but insightful, example from Lewis Carroll was described.

We argued that word problems could be treated as exercises in mathematical modeling, with the basic arithmetical operations providing good models in many situations, approximate models in others, and being inappropriate in yet others. The benefits and problems associated with this radical reconceptualiza-tion are elaborated in Chapter 9, but first, in Chapter 8, we expand our discus-sion of the socio-cultural backgrounds to the use of word problems in mathematics education.

8

Socio-Cultural Contexts of Word Problems

Seven men make 8 bows in 9 days. In how many days do 225 men make 10,000 bows? (from The Shu-shu Chiu-chang of Ch'in Chiu-shao (13th century), Libbrecht, 1973, p. 94)

If a female slave, sixteen years of age, bring thirty-two [nishkas], what will one aged twenty cost? (from Lilavati by Bhaskaracharyya (12th century) translated by Colebrooke, 1967, p. 41)

In this chapter, we offer a wide-ranging discussion of the nature of that "peculiar cultural device" (Lave, 1992, p. 75), the word problem. We begin with a survey of the historical and cultural situatedness of such problems. This survey is followed by analysis of their linguistic structure, followed by discussion of the relationship between text and situations that it putatively describes. We then return to the central role of the immediate cultural environment of students, the mathematics classroom, and how this environment shapes the behavior of students when faced by word problems in school. Finally, we acknowledge the importance of sociopolitical factors, in particular gender and class, that would be appropriate concerns for future research.

Word problems across time and cultures

As illustrated by the above examples, word problems have been found in texts from a very wide range of cultures and going back a long time. There is extensive documentation of the problems themselves, and the texts in which they appear offer fascinating insights into the cultures in which they were written (see, e.g., the examples above). However, there appears to be relatively little discussion in the literature about the overt and implicit purposes of including word problems in mathematics books in different cultures and at different points in history (for an exception see Beckers, 1999).

Just what are word problems *for*? In some cases, it does seem clear enough that the explicit aim was to provide training exercises in practical skills aimed as specific classes of students. For example, according to Beckers (1999), in the Netherlands in the 18th century examples explicitly related to students' future careers as surveyors, merchants, navigators, and so on, were taught in rote fashion. Likewise, Lave (1992), discussing the Treviso arithmetic of 1478, commented that a few master computers accepted merchants-to-be as apprentices and taught them techniques for typical transactions.

Swetz's (1987) analysis at some length of the situatedness of the Treviso arithmetic's instruction within the cultural, historical, mercantile and political contexts of 15th century Venetian life bears out this statement. He concluded that the first part of the book, dealing with computations using the basic operations, was just a preparation for the solving of commercial problems involving such aspects as the Rule of Three, tare and tret, partnership, barter and allegation. He inferred two purposes for the introduction of applications – to teach the techniques of commercial arithmetic, and to provide practice in computational procedures. Swetz traces the Rule of Three (also known as the Merchant's Rule or the Golden Rule) through texts from 1650 BC in Egypt in many cultures of Asia and Europe. Unfortunately (from our point of view), he makes little reference to the question of its appropriateness, though he does point out that "little consideration is ever given to the 'why' of what is taking place" (p. 284) and, at one point, in passing, he refers to its application being "frequently ill-conceived" (p. 225).

Can we speculate about the explicit or implicit intentions of the author of the Treviso arithmetic? It seems implausible that the author was unaware of some of the unreal assumptions that would be needed to justify the answers given in the book (as in the example about building houses discussed by Säljö (1991) – see p. 132). Was the ubiquitous use of proportionality considered as leading to answers that would be reasonable approximations? Or was it recognized that many of the solutions were not realistic, but the student was intended to ignore that for the sake of practicing a particular computational technique? And, if this was the case, did the author have in mind that this technique could be appropriately applied in some situations e.g., currency exchanges or buying goods according to the rule that price is proportional to quantity?

At the stage of history when the Treviso arithmetic was written, mathematics was generally taken uncritically as a description of reality (for example,

Euclidean geometry until the 19th century was regarded as an exact description of space). As Freudenthal (1991, p. 32) puts it:

> Mathematics has always been applied in nature and society, but for a long time it was too tightly entangled with its applications for it to stimulate thinking on the way it is applied and the reason why this works ... money changers, merchants and ointment mixers behaved as if proportionality were a self-evident feature of nature and society ...

> ... Modeling is a modern feature. Until modern times the application of rigorous mathematics to fuzzy nature and environment boiled down to more or less consciously ignoring all of what had appeared to be inessential perturbations spoiling the ideal case.

Besides problems with at least ostensible practical applicability, the Treviso arithmetic also contains such familiar mathematical puzzles as the distance it would take a hound to overtake a hare, given their respective speeds and the distance between them. Swetz comments (pp. 289–290) that these problems:

> ... are not relevant to the commercial world and their origins certainly predate the Treviso writing. They are intellectual exercises and, in a sense, a boast of the power of mathematics. It is almost as if the author is saying to his readers, "Now that you've learned the basics, I'm going to show you something really interesting – some problems you can amaze and puzzle your friends with".

Here we see a very different tradition of "recreational mathematics", with many traditional word problems that have appeared, with variations, across centuries and cultures, as documented by Swetz. The hound and hare problem is a typical example. Swetz describes it (p. 245) as "a standard exercise in European arithmetic books for centuries" having first appeared in Europe in a book of puzzles presented to Charlemagne in about 775 AD. There are many such classic problems, with numerous variants, that have perplexed schoolchildren for centuries. The striking way in which mathematical puzzles, including the oldest known example dating from about 1850 BC, recur across centuries and cultures has also been documented by Wells (1992).

Consider the following problem from *Lilavati*, the 12th century Indian classic by Bhaskaracharyya (Srinivasiengar, 1967, p. 86):

> In the interior of a forest, a number of apes equal to the square of 1/8th of their total number are playing with enthusiasm. The remaining 12 apes are on a hill. The echo of their shrieks by the surrounding hills rouses their fury. What is the total number of apes?

This could hardly be considered as a question that is intrinsically about apes as such. Rather, it is an exercise in carrying out arithmetical procedures, in which the "story" about the apes (apart from the poetic ornamentation) could be considered as a framework functionally equivalent to the use of a letter to indicate a specific but unknown number, and to indicate operations performed on that number which provide the information through which it can be determined. The "story", one might say, provides a grounding for a set of mathematical relationships within an unrealistic, but imaginable, reality (Toom, 1999).

Mathematical puzzles often require the solver to accept conventions. For example, the following is an example of a well-known class of puzzles: "A man is three times as old as his son. In 10 years time, he will only be twice as old. How old are the man and his son now?" Those familiar with such puzzles tacitly assume that age is considered in terms of whole years, yielding the answer "30 years old, 10 years old, respectively" – which would only be exactly true if it is the father and son's common birthday. Question P8 used by Verschaffel et al. (1994) and in subsequent replications (Chapter 2, Table 2.2) is somewhat similar: "Rob was born in 1978. Now it's 1993. How old is he?". Even following the convention that age is given in terms of the last birthday, the appropriate answer to the question is "14 or 15", yet virtually no children gave an answer other than an unqualified 15 (Table 2.5).

In other cases, the solution of a puzzle problem hinges on a "trick" of some kind, as in this example:

> There is a tree of 300 feet high. And under this tree sits a cat who wants to climb to the top. Each day the cat climbs 17 feet, but each night she falls back 12 feet. In how many days will she reach the top? (cited by Dekker, 1996, p. 39)

Versions of this problem, from a 16th century Dutch arithmetic, go back at least as far as 1370 (Wells, 1992, p. 28) – and one of the authors has noticed a version in a recent German textbook series, about a snail climbing out of a well. Many people miss the "trap" in this problem. They realize that each day and night the cat's progress up the tree amounts to 5 feet. Since $300 \div 5 = 60$ the answer 60 is therefore expected. However after 57 days and nights the cat would be 285 feet up the tree and the next day would reach the top. It is worth remarking that teachers and students sometimes refer to the problematic word problems used in our research described in the first part of the book as "trick questions".

The implicit message of such "intellectual exercises", as Swetz (1987, p. 289) describes them, is that realistic considerations should be ignored in the construction of an idealized mathematical model, the essential purpose being to provide exercises in working with mathematical structures. The "story" acts merely as a conceptual vehicle to convey the structural relationships. As expressed by Toom (1999, p. 37) "[t]heir purpose is to convey a *mathematical meaning*, that is the use of suitable concrete objects to represent or reify abstract mathematical notions". Thus, such problems, he argues, perform the function of "mental manipulatives". Responding to the question posed by Gerofsky (1996) as to the purposes of word problems, he commented as follows (p. 36):

> Although Russian educators have been using word problems very productively for a long time, as far as I know they never cared to explain rationally why word problems are so useful, because nobody questioned this in their presence. They were and still remain guided by tradition, experience, intuition and aesthetic criteria, all of which should not be ignored.

and he suggests that "like many other cultural phenomena (fables, for example), word problems have several purposes" (p. 36).

As has been remarked on already, a striking characteristic of many word problems, both puzzles and application problems, noted by many authors (e.g., Swetz, 1987; Wells, 1992), is their continuity across widely different cultures and over many centuries. Säljö (1991, p. 262) commented, with respect to problems from the Treviso arithmetic, that "[they] seem only a little different from tasks than can be found in contemporary text books". Why this stylistic homogeneity? Lave (1992, p. 76) suggested that "perhaps school math includes word problems "because they are there" (and have been for several hundred years)". In similar vein, Gerofsky (1996, p. 43) speculated that perhaps "there is something elemental or common to human experience in these, if we could find it, although perhaps their endurance simply speaks for the incredible conservatism of mathematical tradition".

The stylistic homogeneity is echoed by one particularly striking aspect of the findings reported in Part I of this book, namely the consistency of responses to our set of "problematic" word problems in many countries (Belgium, Germany, Japan, Northern Ireland, Switzerland, and Venezuela) with cognate findings relating to France, Sweden, the Netherlands and the United States. While more focussed cross-cultural studies might be expected to show up interactions between cultural practices of story-telling and school mathematics, what the empirical evidence, together with the textual analysis of word problems, suggests is that there is a remarkable degree of uniformity of practice within mathematics instruction that transcends time and space.

The linguistic nature of word problems

> It takes three men six hours to dig a ditch. How long does it take two men to dig the same ditch? (Traditional, cited by Pimm, 1995, p. 158)

> Suppose a scribe says to thee: "Four overseers have drawn 100 great quadruple hekat of grain, their gangs consisting, respectively, of 12, 8, 6, and 4 men". How much does each overseer receive? (Problem 68, Rhind Mathematical Papyrus, cited by Pimm, 1995, p. 158)

Pimm (1995, p. 158) comments as follows on the above examples:

What in the world might these problems be about? The syntax and style of writing make them stand out as arithmetic problem texts and are unlikely to be mistaken for anything else.

Word problems have been analyzed as "a very special type of text" (Nesher, 1980, p. 41), "a linguistic and literary genre" (Gerofsky, 1996, p. 36). The stereotyped nature of word problems, and manifestations of its effects on students, have been described throughout this book, and analyzed in Chapter 5. As discussed there, one extreme manifestation is that the process of solving word problems often bypasses a genuine attempt to make sense of the situation presented in the text. Thus, "most students perceive word problem solving as a puzzle-like activity with no grounding in factual real-world structures and with no relation to a goal-directed, more authentic activity of mathematization or realistic mathematical modeling" (Reusser & Stebler, 1997a, p. 323).

In many cases, the student presented with a simple word problem text including two numbers uses some superficial cue to decide which of the four basic operations is the one required (Figure 1.3). Such behavior, we have suggested (Chapter 5), is attributable to years of experience in which such a strategy has paid off, and little or no attention has been paid in instruction to developing the ability to discriminate between cases where a simple arithmetical operation offers the appropriate basis for modeling a situation, cases where it is a more or less reasonable approximation, and cases where it is inappropriate.

Nesher (1980) illustrated how this lack of discrimination can be strongly reinforced by the teaching of rules. As an explanation for children's tendency to respond to the question about the temperature of a mixture of one jug of water at 40° F and one jug of water at 80° F by adding 40 + 80 (discussed in Chapter 1, p. 10), she suggested that: "The rule they have learned at school is that when you put things *together*, you add their numbers" (Nesher, 1980, p. 46). Such a rule may be presented explicitly by the teacher to be memorized verbatim. An extreme case, showing how blind adherence to rules that work for routine and stereotyped examples but are not logically sound may be inculcated, is when children may be taught to look for "key words" in the text that will tell them which operation to use. (In Chapter 5 (p. 66), reference was made to a Flemish textbook series in which the key words are printed in a specified color (De Corte et al., 1985)).

In terms of problems involving multiplication and division, the clearest example suggesting rule-bound behavior is the extremely powerful tendency to apply direct proportionality even when it does not appropriately model the situation. This tendency is illustrated by the historical examples at the start of this chapter (particularly the second), by discussion in Chapter 7 (pp. 128–129), and, empirically, by the results for two items used in the research reported in Chapter 2. In the runner item "John's best time to run 100 meters is 17 seconds. How long will it take to run 1 kilometer?" the percentages of students in the various countries who gave the unqualified answer "170 seconds" ranged from 93% to 100% (Table 2.5). For the flask item "This flask is being filled from a tap at a constant rate. If the depth of the water is 4 cm after 10 seconds, how deep will it be after 30 seconds?" [with accompanying figure of cone-shaped flask] the percentages who gave the unqualified answer "12 cm" ranged from 97% to 100% (Table 2.5).

The basic linguistic structure for problems involving (or apparently involving) proportionality includes four quantities (a, b, c, d), of which, in most cases, three are known and one unknown, and an implication that the same relationship links a with b and c with d. In the case of true proportionality, this relationship is a fixed ratio. If a problem matches this general structure, the tendency to evoke direct proportionality can be extremely strong (e.g., De Bock, Verschaffel, Janssens, & Rommelaere, 1999). As a particularly striking example, in one study elementary teachers in training were asked the following: "Sue and Julie were running equally fast around a track. Sue started first. When she had run 9 laps, Julie had run 3 laps. When Julie completed 15 laps, how many laps had Sue run?" (Cramer, Post, & Currier, 1993, p. 159). Of the 33 teachers, 32 gave 15 as the answer, though here the common relationship between Sue and Julie's laps at two points in time is additive rather than multiplicative.

The interpretation of such linguistic structures in terms only of surface structure underlies jokes such as: "If Henry the Eighth had 6 wives, how many did Henry the Fourth have?" More seriously, Lamon (personal communication, 1993) found that many students did not find anything strange about statements such as: "If one orchestra can play a symphony in 40 minutes, two orchestras can play it in 20 minutes." Similarly, Van Lieshout et al. (1997; cf p. 34) found that many children answered the following question as if proportionality was appropriate: "Joris and Pim live in the same house. They bike home together in 8 minutes. How many minutes must Joris bike when he bikes home alone?".

De Bock et al. (1999) argued that the "missing-value problem" formulation (where three numbers are given and a fourth is to be determined) is at least partly responsible for students having such a strong tendency to invoke proportionality inappropriately ("the illusion of proportionality"). They distinguished between two ways of formulating problems, as illustrated by the following examples:

> Farmer Carl needs 8 hours to manure a square piece of land with a side of 200 m. How many hours would he approximately need to manure a square piece of land with a side of 600 m? (Missing-value problem)

> Farmer Carl manured a square piece of land. Tomorrow, he has to manure a square piece of land with a side being three times as big. How much more time would he approximately need to manure this piece of land? (Comparison problem)

It was found that inappropriately proportional responses were much more likely for problems formulated as missing-value problems than for problems formulated as comparison problems. De Bock et al. (1999, p. 247) concluded that "pupils' tendency to apply proportional reasoning in problem situations for which it is not suited is – at least partially – caused by particularities of the problem formulation that pupils learned to associate with proportional reasoning throughout their school career". Thus, both at the level of individual students, and throughout history, we see a non-reflective link built up between the mathematical structure of proportional relationships and a stereotyped linguistic formulation. Such an interpretation is certainly consistent with the results for the runner and flask items, as summarized above.

Generally, mathematical word problems may be considered as a particular linguistic genre, with an associated pragmatic structure (De Corte & Verschaffel, 1985; Kintsch & Greeno, 1985; Nesher, 1980) giving rise to the "word problem game" as discussed in Chapter 5. Given this way of looking at word problems, a number of analysts have considered how constructs from linguistic theory might throw extra light on their nature and use (Gerofsky, 1996; Lave, 1992; Pimm, 1995).

Gerofsky (1996) described word (or story) problems as a special genre of texts with a tripartite structure:

(1)   A "set-up" component, establishing the characters and location of the putative story. (This component is often not essential to the solution of the problem itself).
(2)   An "information" component, which gives the information needed to solve the problem (and sometimes extraneous information as a decoy for the unwary).
(3)   A question.

According to Gerofsky (p. 37):

> ... component 1 of a typical word problem is simply an alibi, the only nod toward "story" in a story problem. It sets up a situation for a group of characters, places and objects that is generally irrelevant to the writing and solving of the arithmetic or algebraic problem embedded in the later components. In fact, too much attention to the story will distract students from the translation task at hand, leading them to consider "extraneous" factors from the story rather than concentrating on extracting variables and operations from the more mathematically-salient components 2 and 3.

Drawing on notions from pragmatics, and more specifically, on Austin's speech act theory (Austin, 1975), Gerofsky points out that understanding a word problem not only involves the understanding of its literal meaning (its locutionary aspect), but also its illocutionary force, namely what performative response it implicitly demands of the hearer. Through enculturation in the genre, students learn to conform to the assumptions of the word problem game (see discussion in Chapter 5). As a result, the illocutionary force of a word problem, as experienced by the learner, could be paraphrased thus (p. 39):

I am to ignore component 1 and any story elements of this problem, use the math we have just learned to transform components 2 and 3 into the correct arithmetic or algebraic form, solve the problem to find the one correct answer, and then check that answer with the correct answer in the back of the book or turn it in for correction by the teacher, who knows the translation and the answer.

From our point of view, this is, of course, a description of *bad* word problem solving – that, firstly, bypasses the situation model and goes directly through some superficial cues from the text to the mathematical model, and, secondly, omits an evaluation of the mathematically derived answer in terms of the initial situation and the situation model (see Figure 1.3).

Elaborating on the idea that children are enculturated into the practice associated with school word problems (for a parallel discussion in relation to word problems in physics see Boote, 1998), Lave (1992, p. 77) commented as follows:

Word problems are not simply a value-free educational or mathematical technology. They provide occasions for talk and activity. There is a discourse of word problems – a set of things everyone knows how to say about word problems or that can be expressed in "word-problemese", issues and questions that come up when people begin to talk about them; and things that are not or cannot be said within this framework. These are stylized narratives about assumed general cultural knowledge that (even) children can be expected to have. They are not about particular children's experiences with the world. The problems themselves are stylized representations of hypothetical experiences – not slices of everyday existence.

The nature of word problems as linguistic entities raises a fundamental question to which we turn in the next section, namely the question of the adequacy of text for providing descriptions of real-world situations.

The relationship of text to reality

In terms of the relationship between aspects of reality and the mathematical structures used to model them, discussed in general terms in Chapter 7, specific questions are raised by the presentation of a situation in a word problem through text (though not necessarily, as noted in the introduction (p. x), exclusively text).

One standard justification for word problems is that they show, albeit in simplified form, how mathematics can be used to model real-life situations (see p. xi). The validity of this claim has been forcefully questioned. For example, Nesher (1980, p. 42) stated that "[word problems] must be recognized as a very special type of text, whose interpretation is shaped by the language game of arithmetic instruction, and *cannot be considered as a text describing real life situations"* (emphasis added). While it is true that many word problems are impoverished or even inaccurate descriptions of real-life situations, the italicized phrase looks like an overstatement. If a text that refers to several people eating in a restaurant, gives information on how much each person's meal and drinks cost, the amount of tax and tips, and asks what the total bill comes to is *not* a text describing a real life situation, *what would be?*

Lave's characterization of word problems as "stylised representations of hypothetical experiences" seems more balanced, and is applicable generally to learning and instruction. For example, a law student may be presented with the outline of a case and asked to make and justify some judgment. A medical student may be presented with some details of a case and asked to make a diagnosis. A person learning to play bridge or chess may study hands or positions presented in a book, as an alternative to learning by playing. In fact "stylised representations of hypothetical experiences" may be used in any form of training and have obvious advantages – they are cheaper in money and time, they can be chosen with specific pedagogical goals in mind and they avoid the consequences of "doing the real thing" particularly by inexperienced learners (think of an airline pilot training on a simulator or a surgeon in training).

Gerofsky (1996, p. 40) commented that: "...word problems propose hypothetical situations with certain given conditions and ask for hypothetical answers". But does it matter whether or not the situation is hypothetical? In the case of the bridge or chess player, does it matter whether the hand or position, respectively, once occurred in an actual game or has been constructed by the author for the purposes of teaching? In the restaurant bill word problem, does it matter if the details were made up or a record of an actual meal as long as the essential elements are taken into account, the given numbers are realistic, and the question asked is meaningful? Where it could have an effect is in terms of sending the message that mathematics does apply to real experience. Contrast a word problem of this sort presented in a textbook with a similar problem presented by the teacher taking out of her pocket the bill from the meal she had with friends the night before and asking a question about it.

In terms of her suggested three-part structure for word problems, Gerofsky (1996, p. 37, and see p. 146) states that the "set-up" component of a word problem establishes "a situation for a group of characters, places and objects that is generally irrelevant to the writing and solving of the arithmetic or algebraic problem embedded in the later components". This is certainly true in the sense that we have no difficulty supplying an answer for a problem such as "A twratrw has 3 grodi slogwters. A nitata gives the twratrw 6 more grodi slogwters. How many grodi slogwters does the twratrw have now?". This example neatly illustrates the difference between the textual elements that convey the gist of the underlying situation (having, giving, and the additive change structure) and the incidental aspects of who has and who gives what.

However, it is not clear what point Gerofsky is making when she goes on to notate a word problem as follows:

> Every year (but it never happened), Stella (there is no Stella) rents a craft table at a local fun fair (which does not exist). She has a deal for anyone who buys more than one sweater (we know this to be false). She reduces the price of each additional sweater (and there are no sweaters) by 10% of the price of the previous sweater that the person bought (and there are no people, or sweaters, or prices …)

A similar treatment of the first sentence of Kafka's *Metamorphosis* would read as follows: "As Gregor Samsa (there is no Gregor Samsa) awoke one morning (there was no such morning) from uneasy dreams (there were none) he found himself transformed in his bed (there is no bed) into a gigantic insect (there is no such insect)".

As already illustrated by reference to a range of instructional settings, Lave's (1992, p. 77) characterization of word problems as "stylised representations of hypothetical experiences" may be applied to more than word problems. Eco (1994, p. 2) quotes the start of an Italian folk-tale:

> A king fell ill and was told by his doctors, "Majesty, if you want to get well, you'll have to obtain one of the ogre's feathers. That will not be easy, since the ogre eats every human he sees".

and the comments on it by Calvino (1988) that "not a word is said about what illness the king was suffering from, or why on earth an ogre should have feathers". Eco (1994, p. 3) declares that "every text is a lazy machine asking the reader to do some of its work" and goes on to quote another story, from Schank (1982, p. 21):

> John loved Mary but she didn't want to marry him. One day, a dragon stole Mary from the castle. John got on top of his horse and killed the dragon. Mary agreed to marry him. They lived happily ever after.

Schank used this very stark narrative to investigate what children understand when they read. On the basis of one interview of a three-year-old child, Eco (1994, p. 6) suggests that "the girl's knowledge of the world included the fact that dragons breathe flames from their nostrils but not that you can yield, out of gratitude or admiration, to a love you do not reciprocate". This echoes Lave's (1992, p. 77) description of word problems as "stylized narratives about assumed general cultural knowledge that (even) children can be expected to have". Consider again the problem about the bill in a restaurant. There is general cultural knowledge about the typical course of events when people go to a restaurant for a meal – Schank and Abelson (1977) referred to this knowledge as a "restaurant script". In interpreting a story about a restaurant this script makes possible inferences beyond what is explicitly stated in the text. For example, part of the script is the default assumption that the cost of the meal includes the added together costs of the items ordered by the consumers as listed on the menu, as well as taking into account the more context-specific aspects of tax and tipping.

In the wake of increasing emphasis on teaching applications of mathematics (e.g., de Lange, 1996; Keitel, 1996), attempts by textbook authors and test constructors to present students with "realistic" word problems often seem ill-conceived and have been extensively criticized (e.g., Cooper, 1992; Dowling, 1996). For example, Cooper (1992) analyzed an item devised for national testing in the United Kingdom (see earlier discussion on p. 69). The item has a diagram showing a sign in an office block, saying that "This lift can carry up to 14 people" and a statement that, in the morning rush, 269 people want to use it. The question asked is: "How many times must it go up?". The marking scheme specifies that an answer of 19 or 19.2 should not be accepted. (Presumably the test setters were aware of the research on the buses problem (p. 6) and related examples). Cooper (1992, pp. 234–5) commented as follows:

> Now, does this reduce to 269 ÷ 14 rounded to the next largest whole number, as the Marking Scheme supplied with the tests implies? Well, possibly, if the child realizes that it should not be treated as a real "real life" problem. For example, the child has to assume, at least implicitly, in order to find the "right"answer:
>
> • That the lift is always full, except for the last trip ...
>
> • That nobody, in this "morning rush" for the lift, gets fed up and decides to use the stairs
>
> • That everyone involved uses the normal space associated with a person in a lift. Nobody here, amongst the 269, is, for example, in a wheelchair.
>
> In other words, the child has to ignore the reference to the "real world" in order to succeed at the piece of arithmetic, except for avoiding the "catch" of merely dividing 269 by 14 and giving the "unreal" answer of 19 or 19.2 ...
>
> ... Some reference to the "real" must be made, but not too much.

(Note that in our categorization of responses to the similar buses item in the studies reported in Chapter 2, a response raising any question about necessary assumptions would have been rated as RR).

This example raises a number of fundamental points. In particular: How is a student expected to judge how much complexity to assume? This is an extremely important point, but not, perhaps, particularly well illustrated by this example. The answer 20 allows for leeway (for wheelchairs, for example) – it's an appropriate answer provided that, in most cases, the number in the lift is at least close to 14. From our perspective, moreover, the understanding why 20 may be considered a better answer than 19 or 19.2 is, emphatically, not simply a matter of a "catch". Another aspect that might be queried is "Why would anybody *want* to know the answer to the question asked?".

Another question analyzed by Cooper (1992, p. 235) concerns a choice between two players for a basketball team. The scoring record for each player in 6 matches is shown, and the student is expected to justify the choice of one player or the other, making reference to both the mean and the range of the players' scores: "It doesn't matter which player you choose but you must use the mean as well as the range to explain which player is better". The intent of the examiner seems clear. Both "Karen" and "Jo" have averaged 12 points per match (note, in passing, the use of "neat" numbers) but Jo's performance is much more variable (as reflected in a range of 28 points, as opposed to Karen's range of 4 points). It seems clear that the examiner wants to gauge if the student understands how variation is important as well as average. The item has the virtue, at least, that it acknowledges the possibility of more than one answer, with supporting arguments for each.

In contrast to the lift situation, where it can be contended that reasonable assumptions can be made, this example seems grossly unrealistic in that essential information that would be used in a real scenario is missing. The decision is to be based on the result of processing the numerical data provided. In an actual situation involving the selection of a player, the decision would be based on much more complex human aspects, such as the temperaments of the players, the assessed relative strength of the opposition (if they are considered stronger, then a player who is inconsistent but capable of "rising to the occasion" might be preferred to one who is consistent but rarely outstanding, for example), and so on. These flaws in the question illustrate the distinction between closed and open tasks. School problems and, even more so, assessment predominantly are presented as informationally closed – that is, the student cannot seek further information as appropriate, which would generally be possible, and typically exploited, in an authentic problem-solving task (see p. 72).

To sum up, recent critiques have called – appropriately, in our view – for a re-analysis of the rationale for word problems in the curriculum. Questions have been raised, in particular, about the adequacy of word problems to portray authentically real-life situations in all their complexity. Although these criticisms carry a great deal of weight, we have suggested that at times they have been exaggerated, and our contention is that word problems *can* constitute a form of text that relates to reality (actual, hypothetical, or imagined experience) and allows students to draw on general cultural knowledge beyond the text, and that such word problems – if improved in numerous ways – continue to have a useful function in demonstrating the applicability of mathematics, as discussed in the next chapter.

Word problems in the social context of the mathematics classroom

As has been documented earlier and emphasized throughout (particularly in Chapter 5), in the course of our research and delving in the literature, it soon became apparent that any explanation of the students' observed behavior framed solely in terms of cognitive factors is inadequate, and that, rather, the explanation should be sought primarily in the nature of the learning environments experienced by the students. In particular, it is appropriate to invoke Brousseau's (1984, 1990, 1997) concept of "didactical contract" to describe the, largely implicit, rules of the classroom, the forms of behavior of the teacher expected by the pupils and vice versa. The behavior of the children, in apparently ignoring realistic constraints may be attributed to their adherence to the didactical contract, as discussed in Chapter 5 (pp. 60–62). By analogy with the didactical contract, Greer (1997) suggested that, in a research setting, there is a (usually implicit) "experimental contract". Arguably, the experimental contract of the early experiments using pencil-and-paper tests as described in Chapters 1 and 2 was similar to the didactical contract in traditional mathematics classrooms, and this provides a possible explanation for the results obtained. Some support for this supposition comes from the much higher rate of realistic responses obtained in the experimental conditions reported in Chapter 4 where the experimental contract was radically changed (as opposed to the relatively superficial changes in the experiments reported in Chapter 3, which produced only small effects).

Along the same lines, several writers have pointed out that, although superficially there appears to be "suspension of sense-making" (Schoenfeld, 1991) on the part of the children, on closer examination their behavior can be construed as sense-making of a different sort. Schoenfeld himself openly expressed his conversion to this point of view as follows:

> Taking the stance of the Western Rationalist trained in mathematics, I characterized student behavior on the NAEP bussing problem [see p. 6] – a violation of my particular epistemology – as a violation of sense-making. As I have been admonished, however, such behavior is sense-making of the deepest kind. In the context of schooling, such behavior represents the construction of a set of behaviors that results in praise for good performance, minimal conflict, fitting in socially etc. What could be more sensible than that? The problem, then, is that the same behavior that is sensible in one context (schooling as an institution) may violate the protocols of sense-making in another (the culture of mathematics and mathematicians). (Schoenfeld, 1991, p. 340)

(For similar views, see, e.g., Gravemeijer, 1997; Reusser & Stebler, 1997a; Selter, 1994.)

What has just been said about the students applies, *mutatis mutandis*, to the teachers, whose behavior is covertly controlled by the didactical contract just as much as that of their students, and whose behavior is equally rational in terms of the system within which they operate (see discussion on pp. 60–62). Empirical evidence supporting this assertion was found in the study with teachers-to-be reported in Chapter 5 (pp. 73–82).

The inference of underlying cultural rules from observed behavior in a particular setting, namely the mathematics classroom, suggests that the phenomena might be explored in ways analogous to those used by anthropologists. Indeed, in one of the first examples of suspension of sense-making that came to our notice (introduced on p. 10), Davis (1989, p. 144) suggested of the student who used scissors in order to share a balloon that his behavior showed that he was adapting appropriately to what Davis called the "peculiar tribal culture of the American classroom" (as we have seen, it is hardly necessary to restrict this quotation to the American classroom). Likewise, Schoenfeld (1987b, p. 38) spoke of playing the role of amateur anthropologist.

Jacob (1997, p.3) has argued that, until comparatively recently, a unified approach to educational innovation that considers the interaction between context and cognition has been lacking:

> Mainstream educational developers and researchers, using positivist lenses, have focused primarily on cognitive issues ... and have largely ignored context. They have had cognition without context. Educational anthropologists, using interpretivist lenses and generally operating outside the mainstream, have focused primarily on the contexts of education, and usually have not addressed "taught cognitive learning". They have had context without cognition.

However, in recent times the picture has radically changed. The interaction between mathematical cognition and context has been demonstrated in a variety of cultural contexts, early examples being:

- Scribner's (1984) study of the arithmetic strategies used by dairy workers making decisions about prices and quantities in the course of their daily work.
- Studies of the arithmetical skills of street vendors in Brazil (Carraher et al., 1985).
- Lave's (1988) study of arithmetical processes of supermarket shoppers.

Such studies contributed significantly to the prominence achieved in cognitive psychology by the concept of "situated cognition" (Brown, Collins, & Duguid, 1989; Kirschner & Whitson, 1997). A highly prominent theme, both for the studies of mathematical cognition, and of cognitive psychology, has been the contrast between the forms of learning that take place in and out of school (e.g., Resnick, 1987). Moreover, many studies, including those cited above, have shown there is a wide gulf between the mathematics children learn in school and the reality of their experiences outside school in that mathematics learnt in school is typically not used outside of school. Our studies show a converse aspect, namely that children do not bring everyday knowledge to bear when solving word problems – or rather, the context of school mathematics conditions them not to do so.

While the term "situated cognition" has usually borne connotations of work and other everyday activities outside of school, Lave (1992) characterized mathematics in school as itself a form of situated cognition, explaining that the term "situated" doesn't necessarily mean that the activity is concrete and specific, but rather that: "It implies that a given social practice is multiply interconnected with other aspects of ongoing social processes in activity systems at many levels of particularity and generality" (p. 84).

From this perspective, then (p. 81):

> ... math in school *is* situated practice: school is the site of children's everyday activity. If school activities differ from the activities children and adults engage in elsewhere, the view of schooling must be revised accordingly; it is a site of specialized everyday activity – not a privileged site where universal knowledge is transmitted.

(For a detailed analysis of children's mathematical behavior in contrasting instructional environments, using situated cognition as a theoretical framework, see Boaler, 1997b).

Up to this point, the situatedness of school mathematics has been discussed in terms of its negative connotations, specifically in relation to how word problems have traditionally been, and for the most part still are, taught. But characterizing mathematics in school as situated practice does not rule out the possibility of changing that practice. A more optimistic view was expressed by Cobb et al. (1993, p. 96) in describing changes in the social practice of students and teachers in their study:

> We question [the] metaphor of either students or teachers being *embedded* or *included* in a social practice. Such metaphors tend to reify social practices, whereas we believe that they do not exist apart from and are interactively constituted by the actions of actively interpreting individuals ... By making this point, we are attempting to avoid any tendency that subordinates the individual to the social and loses sight of the reflexive relation between the two.

In the final chapter, we will reflect on ways in which the practice of school mathematics could be changed for the better (in our opinion), with particular reference to the examples of design experiments discussed in Chapter 6.

Sociopolitical factors

There are several important sociopolitical perspectives on the practice of word problems in schools that we must acknowledge have not been addressed in our research, that certainly should be foci for future research, and that demand at least brief discussion here.

The most obvious is consideration of gender effects. A common theme is that performance on application problems is affected by the degree to which they are aligned with student interests (e.g. Anand & Ross, 1987), differences in which include those that are gender-related (Goodell & Parker, in press). That the relationship between context and performance is not simple is illustrated by Boaler's (1994) finding that female students performed relatively worse on an item relating to fashion than on an item relating to football. The explanation put forward by Boaler was that "girls are more likely than boys to underachieve in contexts which present real world variables but do not allow the variables to be taken into account" (Boaler, 1994, p. 551). Her discussion of the particular item relating to fashion echoes the criticisms considered earlier, summed up by Cooper's (1992, p. 235) conclusion that "Some reference to the 'real' must be made, but not too much". More generally, Boaler's example of how girls may be disadvantaged fits within critiques of mathematics and mathematics education of excluding "women's ways of knowing" (Becker, 1995; Belensky, Clinchy, Goldberger, & Tarule, 1985; Boaler, 1997a; Damarin, 1995).

A second differential effect to which Boaler (1997b), amongst others, has drawn attention, is that of the nature of the instruction. In her detailed study of the highly contrasting mathematics instruction in two schools, she found that the female students at Amber Hill school articulately explained that their belief in the value of understanding was incompatible with the traditional, textbook based style of instruction and practices which "encouraged the students to locate their mathematical knowledge within the four walls of their mathematics classrooms" (Boaler, 1997b, p. 144). By contrast, the male students at Amber Hill "were not happy, but they were able to play the game, to abandon their desire for understanding and to race through questions at a high speed" (p. 119). At the other school, Phoenix Park, a radically different teaching style was used, based on open-ended projects, considerable freedom for the students, and less clear boundaries between school mathematics and the world outside; in this environment, female students outperformed males.

Parallel arguments and findings about the differential effects of spurious realism in application problems in relation to social class have been presented by Cooper (Cooper, 1992; Cooper & Dunne, 1998a, 1998b; Cooper, Dunne, & Rodgers, 1997). In particular, drawing on a body of sociological theory, Cooper (1992, p. 241) suggests that, when approaching contextualized test items:

> Working-class children might be less likely to step outside an "everyday" frame of reference. Middle-class children on the other hand, might be more ready to spot and/or accept the rules of this testing exercise, and be more ready to switch to the required approach.

In an analysis of performance on national tests taken by 10–11-year-olds in England, Cooper and Dunne (1998a) showed that working-class children (and also female students) performed, in relative terms, much worse on "realistically" contextualized test items than on "esoteric" items with no ostensible link to everyday contexts.

The work of Cooper and others is particularly relevant in relation to the effects of assessment, both in terms of how forms of assessment may produce differential gender and class effects and in terms of the dilemma facing teachers who, if they want to maximize scores in tests based on traditional forms of assessment, must warn students to "beware references to the practical, real and everyday world" (Cooper, 1992, p. 242). Another component of the instructional environment, namely textbooks, has been analyzed in detail in relation to the social class of targeted students by Dowling (1991, 1996, 1998).

In terms of cultural/ethnic differences, we have remarked on the fact that experimental findings have shown remarkable consistency across many countries (Table 2.5). This consistency of results may be interpreted within the theoretical framework of situated cognition as reflecting a degree of uniformity in the instructional environments of mathematics education. As educational systems work towards more differentiated curricula, it may be that views on the relationship between reality and mathematics, and specifically on the degree to which application problems should be culturally appropriate and take into account real-world considerations, may lead to more diversity. More generally, as it becomes more widely accepted that mathematics and mathematics education are inherently political (D'Ambrosio, 1999; Mukhopadhyay & Greer, in press), there are calls for application problems to deal with serious and controversial sociopolitical topics, notably by Frankenstein (1989, 1996).

Summary

In this chapter, we have considered word problems from several perspectives.

Examining the historical record may be enlightening in several respects – in showing parallels with conceptual problems of contemporary students (for example, in relation to proportional reasoning), in describing the evolution of practices of schooling, and in illustrating the situatedness of cognition in relation to instructional environments. Our review also strongly suggests that word problems have shown remarkable homogeneity over centuries and across cultures, an observation that deserves further analysis.

Next, we considered word problems as linguistic objects forming "a linguistic and literary genre" (Gerofsky, 1996, p. 36). Bringing to bear tools from branches of linguistics, and comparing word problems with other linguistic objects, allow us to lay bare the implicit rules that govern the interpretation of text in word problems, rules that, through familiarity, are normally unexamined. Such rules, we have argued, are implicated in students often making superficial connections between word problem texts and the situations they are intended to describe rather than seriously attempting to mathematically model those situations.

A limitation of the research we have reported is that, in describing the behavior of students solving word problems, relatively little attention has been paid to individual differences, either through detailed studies of individual students or through consideration of important variables, notably gender and social class. Such limitations offer clear pointers for future research.

9

Reconceptualizing the Role of Word Problems

It was a lesson under the heading of "ratio and proportion" and the teacher told me that she wanted to approach the mathematical concepts in a practical way. So she offered the following question: "Somebody is going to have his room painted. From the painter's samples he chooses an orange colour which is composed of two tins of red paint and one-and-a-half tins of yellow paint per square meter. The walls of his room measure 48 square meters altogether. How many tins of red and yellow are needed to paint the room the same orange as on the sample?" The problem seemed quite clear and pupils started to calculate using proportional relationships. But there was one boy who said: "My father is a painter and so I know that, if we just do it by calculating, the colour of the room will not look like the sample. We cannot calculate as we did, it is a wrong method!" In my imagination I foresaw a fascinating discussion starting about the use of simplified mathematical models in social practice and their limited value in more complex problems (here the intensifying effect of the reflection of light), but the teacher answered: "Sorry, my dear, we are doing ratio and proportion". (Keitel, 1989, p. 7)

In this final chapter, we present our views on the implications of the results of our research and analysis, as described in Parts 1 and 2 of this book, and the further reflections and analysis of the preceding two chapters. Starting from the standpoint that the need to do something about word problems has been demonstrated, we consider the various classes of word problems, with different natures and purposes, and the continuing role they should have in school mathematics. For this continuing role to be justified, however, major reforms are needed in the instructional design of word problems, and associated assessment techniques.

In line with a recurring theme throughout this book, we then suggest that reconceptualizing word problems as an exercise in mathematical modeling offers many advantages. A radical change of this nature would have to be implemented in the context of changes in classroom culture that are consistent with the recommendations of many mathematics educators. In particular, we argue that modeling, in several respects, is inherently a social process. Such an emphasis fits well with contemporary views on reconstructing the mathematics classroom as an environment for collective and negotiated sense-making.

While acknowledging that such changes involve great difficulties and imply the expenditure of a considerable amount of human resource, we argue that the cost of maintaining the status quo, in terms of the long-term effects on how students perceive and do mathematics, would be much greater.

The need for change

It is very widely agreed that something needs to be done about word problems. The findings reported in this book show that children (and teachers) are induced to connive in a form of suspension of sense-making that runs counter to most contemporary statements of goals for mathematics education.

A first necessary stage is to examine and problematise the largely unexamined assumptions about the rationale for including word problems in mathematics education – as Gerofsky (1996, p. 43) says: "to think in new ways about the nature and purpose of word problems, about their inherent oddness and contradictions, and about our rationale for using them in school mathematics programs, rather than simply, unthinkingly visiting them upon future generations of schoolchildren".

The historical review presented in Chapter 8 revealed that word problems, in terms of the types of problems and their linguistic presentation, have shown a remarkable continuity across cultures and many centuries. Lave (1992) and Gerofsky (1996) suggest that there is considerable inertia in the educational system whereby practice is passed on from generation to generation without examination of the reasons, some of which may have been valid in the past but have become obsolete. In particular, in the past, word problems were used as part of targeted vocational training for selected groups. As Lave (1992, p. 74), discussing the Treviso Arithmetic (Swetz, 1987), put it: "In this socially organized math practice the apprentices learned from the master computers the math they needed to carry out typical business transactions", a situation which no longer applies today when all children are expected to learn a wide range of mathematics supposed to be generally applicable.

A major part of the problem, as we see it, is the epistemology that is still largely predominant among people in general, mathematics teachers, and, to some extent, mathematicians, whereby mathematical statements are regarded as the touchstone of certain and indubitable truth. As argued in Chapter 7, the difficulty lies in extending this absolute truth from formal statements of pure mathematics to statements about aspects of reality. In particular, throughout history, and throughout the history of mathematics education, the link between word problems that are ostensibly about real-world situations and simple mathematical structures has been unmindfully regarded as obvious and non-problematic. Overwhelmingly, the teaching of word problems has both reflected and reinforced such a view, and continues to do so. There is a failure to develop critical discrimination between cases when a particular mathematization is precise, when it is a more or less good approximation, and when it is inappropriate.

The study of teachers-in-training reported in Chapter 5 suggests that their view of word problems (and, by implication, that of their instructors) is consistent with this contention. Commenting on this study, Gravemeijer (1997, p. 391) pointed out fundamental differences in beliefs about word problems held by the researchers (i.e. those represented in Greer & Verschaffel, 1997) and the teachers. Specifically, he suggests that the teachers believe that the activation of realistic context-based considerations should be discouraged rather than stimulated. We agree with Gravemeijer (1997, p. 391) that a fundamental value judgment underlies the reasons for using word problems:

> The researchers relate word problems to problem solving and applications. The student teachers (and teachers in general, probably) see another role for word problems. It seems likely that for them the primary role of word problems is that of exercises. In such a conception, word problems are nothing more, and nothing less (!) than decorated exercises in the four basic operations.

The implication, as he sees it, is that opting for sense-making rather than computational proficiency as the major reason for using word problems implies a radical change in the goals of mathematics education.

In the following sections, we make suggestions as to how the teaching of word problems could be reformed by considering, in turn, design improvements in the nature of the problems to be used, the form and nature of methods of assessment, the potential of the modeling perspective, and, most importantly, the nature of the instructional environment.

Improving the quality of word problems

As argued in Chapter 8, word problems differ in terms of their implicit nature and their purposes. Here we suggest a categorization into four classes.

First, there are word problems describing real-world situations which are "cleanly" and appropriately modeled by applications of the basic arithmetical operations. The S-problems used in our research (Table 2.2) are typical examples. As has been extensively documented and researched for both addition and subtraction and multiplication and division (e.g., Fuson, 1992; Greer, 1992; Riley et al., 1983, Staub & Reusser, 1995; Verschaffel & De Corte, 1993, 1997a) the variety of such situations is very wide, and the difficulty of understanding the relationship of the situation to the corresponding arithmetical operation differs widely. For example, for young children, problems involving additive comparisons are relative difficult, and for older children, Cartesian product is a less obvious application for multiplication. Moreover, there are considerable difficulties for children in extending additive concepts to directed numbers and multiplicative concepts to non-whole numbers. Word problems in this class constitute an essential component in developing an understanding of the additive and multiplicative conceptual fields, a process that extends over many years (Vergnaud, 1996). The problem, as we have argued throughout this book, is when they come to dominate children's instruction to the extent that the assumption is created that any word problem can be solved by applying one of the four basic arithmetical operations to the two numbers mentioned in the problem.

Second, there are word problems describing real-world situations that are more complex than those in the first class. For some such situations, the application of a single arithmetic operation may still lie at the core of the mathematical model of the situation, but that model needs to be refined in some way. A well-known example is division with remainder problems, where the computation involved is division, but the result needs to be interpreted in the light of the situation described (see the discussion of the first TLU used by Verschaffel and De Corte (1997b), Chapter 6 and Figures 6.1–6.4). Another set of examples have subtraction as the core operation, but complications in that the result of the computation may need to be adjusted by ± 1, depending on the situation, as covered in the third TLU of the teaching experiment carried out by Verschaffel and De Corte (1997b) and reported in Chapter 6 (pp. 89–96) (see also Verschaffel, De Corte, & Vierstraete, 1999). In other situations, such as the problem involving a flask of diminishing cross-section, a precise model of the situation involves more complex mathematics, including, in that case, calculating the volume of the frustum of a cone. In yet other cases, the application of an arithmetic operation may be done on the basis of simplifying assumptions – an idealization of the situation – to yield an answer that must be interpreted as an approximation. Speed/distance/time problems are usually of this nature, if one takes seriously the realistic considerations (e.g., nobody drives a car at a constant speed for 4 hours) but they are typically treated as if they are problems of the first sort. As discussed in detail below (and adumbrated in Chapter 7), our proposal is that both of the above classes of word problems should be considered within a framework of mathematical modeling.

The third class is of word problems that describe situations that are clearly unrealistic or even fantastical, but imaginable. It includes what Pollak (1969) called "problems of whimsy" in which the unreality is apparent and would generally be recognized as such, calling for a willing suspension of disbelief. An example provided by Pollak (1969, p. 397) is the following:

> Two bees working together can gather nectar from 100 hollyhock blossoms in 30 minutes. Assuming that each bee works the standard eight-hour day, five days a week, how many blossoms do these bees gather nectar from in a summer season of fifteen weeks?

De Lange (1996, p. 67) comments on this example that it requires application of mathematics in "an applied (but not so real) situation, using well defined computational tools". The introduction of unusual or bizarre contexts, and possibly fantasy, may have some motivational force in making the problem interesting.

Fourth, there are those word problems that may be described as puzzles of various sorts. The distinction between puzzles and problems of whimsy is not clear-cut, but generally they are less amenable to standard solution techniques and often they involve some subtle insight to solve. Examples were given in Chapter 8, many of which originated in antiquity and have reappeared throughout history and across cultures in numerous variations (Swetz, 1987; Wells, 1992). The following comes from Greece (c. 500 AD):

> I am a brazen lion; my spouts are my two eyes, my mouth and the flat of my right foot. My right eye fills a jar in two days, my left eye in three, and my foot in four. My mouth is capable of filling it in six hours; tell me how long all four together will take to fill it? (Cited in Wells, 1992, p. 11)

The history of such puzzles, and the continuing popularity of "recreational mathematics" testify to the importance of intellectual play in mathematics (Toom, 1999).

Problems of whimsy and puzzles have in common that they are usually assumed to map cleanly onto mathematical structures (in the case of puzzles there may be implicit conventions necessary to make this possible). For both types, the "story" acts as a vehicle to convey structural relationships. Both types have a legitimate place in mathematics instruction as a means of promoting thinking, problem solving and conceptual development. The use of non-realistic word problems may be justified on two major grounds, as argued by Toom (1999) (and consistently with the Realistic Mathematics Education approach of Freudenthal and his followers (e.g. Freudenthal, 1991; Gravemeijer, 1994; Treffers, 1987)). Firstly, as discussed in Chapter 8, they can act as "mental manipulatives" providing imagistic (in a broad sense) support for working with mathematical structures. Secondly, they have a legitimate place in terms of children's creativity and imagination. Toom (1999, p. 38) ridicules the notion that every word problem must relate to some present or future real-world situation in the life of the student "as if Aesop's fable is useful only for those who have the chance in some already-pre-perceived future to perch on a tree branch with a piece of cheese in their mouths".

While hardly constituting a class of word problems, there is a variety of forms of wit that convey various points about the nature of the genre – and could be used instructionally for this purpose, as argued later. Examples were given in Chapter 5 (p. 66) of jokes that illustrate the difference between word problems and problem solving in real life, such as the one about six birds sitting in a tree and two being shot by a hunter. There are also nonsensical questions with the surface structure of word problems that could also be used for instructional purposes, such as "If Henry the Eighth had 6 wives, how many did Henry the Fourth have?".

In making these distinctions between different types of word problem, a rough parallel could be drawn with the literary arts, in which we can distinguish types of books, plays and films such as fictitious works intended to illuminate some aspect of human experience in a simplified or symbolic form, including parables (Pimm, 1995, p. 161) and fables (Toom, 1999); works intended, with some degree of verisimilitude, to tell a "true story" about events that happened – or might have happened; works of fantasy, such as nursery rhymes, fairy tales, some kinds of science fiction; works that "play" with the medium (verbal or visual) in a puzzle-like way and with the reader's assumed knowledge of the conventions of different genres (Eco, 1994).

There remains a place in the curriculum for all the forms of word problem considered above. As Staub and Reusser (1995, p. 302) commented:

> Because word problems have clear final processing goals and because a successful solution depends on thoroughly understanding the situations described in a text, these problems provide excellent opportunities to explicitly apply world knowledge, discourse and language knowledge, as well as arithmetic knowledge.

However, more attention needs to be given to design issues:

> From an instructional point of view, we need to inquire about the goals, purposes, and plans that are related to the presentational structure of specific word problems. We think that educational psychology can help improve instructional design by contributing to educators' knowledge about how this variation of presentational structures affects both problem difficult and solution strategies. This instructional knowledge could provide criteria for selecting or generating problem texts whose instructional objectives are to foster students' specific knowledge and skills in comprehending textually presented situations that are to be mathematized. Thus, they contribute to students becoming flexible discourse and problem comprehenders. (Staub & Reusser, 1995, p. 302)

Criticisms, and recommendations for the improvement of word problems, are not new (e.g., Nesher, 1980; Puchalska & Semadeni, 1987; Reusser, 1988). In

Chapter 5, a thoroughgoing critique of the stereotyped and artificial nature of word problems in mathematics lessons and assessments was presented. In terms of the eight characteristics listed there (pp. 67–70) that are considered to contribute to the harmful effects of the impoverished diet typically fed to students, the following recommendations may be made.

First, it would be very easy to break up the expectation that any word problem can be solved by the application of one of the four basic arithmetic operations to the two numbers mentioned in the problem. From an early stage, easily understood counter-examples could be introduced (the P-problems used in our research (Table 2.2) provide examples), with discussion explicitly aimed at discrimination learning (cf Barron et al., 1998).

A particular example concerns those cases where the "core operation" is division, but the result of the computation must be interpreted depending on the nature of the situation. Streefland (1988, p. 81) describes an exercise in which students were asked to invent stories for the division 6394 ÷ 12 for which the appropriate answer would be 532, 533, 532 remainder 10, and so on (quoted at the head of Chapter 6, p. 85). In the teaching experiment by Verschaffel and De Corte (1997b) reported in Chapter 6 (pp. 86–96), the first TLU addressed exactly this point.

Another variation from the single-operation pattern is to introduce problems involving two or more operations, while complex problems (such as the CTGV's Jasper adventures described in Chapter 6 (pp. 108–112) target the belief that any problem can be solved within 5 minutes.

Second, insofar as problems involving a single operation are presented, it would be a straightforward matter, and the most minimal of changes, to redesign textbooks so that the same operation is not repeated across a block of items, or under a heading that tells the students what to do! Counter-examples, with associated discussion, could be used to show that specific key words are not an infallible guide to choosing an operation. In this respect, a potentially effective exercise would be to have students construct their own examples in which standard key words offer misleading cues as to the appropriate operation.

Third, problems can be varied by presenting some with superfluous, misleading, or missing data, targeting the assumption that all the numbers in the problem statement, and no others, should be fed into the calculations. One approach is to present the students with rich information from which they need to decide what to select. For example, Lesh and Lamon (1992a) provided children with old and current mail-order catalogues and asked them to estimate the rate of inflation on the basis of the changes in prices. The Jasper adventures also require students to select what data to use from an informationally very rich background. In the case of missing data, students may be expected to find out the appropriate information or make a reasonable estimate (as in the case of the polar bear example discussed on pp. 114–115). A nice example of a problem with no data as such is that provided by Pollak (1987, p. 263), namely how many items should be allowed in the express check-out at a supermarket. There is no obvious way to model this problem (although he asserts that there is at least one beautiful formulation for it).

Fourth, much more care needs to be devoted when creating problems that are supposed to be realistic application problems, particularly in the context of written assessment that, typically, is closed in terms of information. In relation to helping teachers develop a sense for realism, Lesh and Lamon (1992a, p. 45) discuss a problem about softball (a variant for baseball for young players) that appeared in the Curriculum and Evaluation Standards (National Council of Teachers of Mathematics, 1989). In this problem a table gives data on the last 100 times batting for a player named Joan (9 home runs, 2 triples, 16 doubles, and so on) and asks questions such as "What is the probability that Joan will get a home run?" next time she comes up to bat. (Note, in passing, the neatness of the 100 that, doubtless with good intention, makes the computation of the probability easy). The critique of a teacher (herself a softball player) included the following:

> On the surface, this problem appears to be embedded in a real-world situation: Joan is coming in to bat, and the problem description gives some data about her prior performance. But, in a real situation, it wouldn't be sensible for someone (other than a math teacher) to want to know the answer to the questions as they are stated ... In fact, simply computing this "probability" using the intended rule depends on ignoring common sense and/or practical experience. In reality, the probability depends on who is pitching ... on field conditions or the weather, and on a lot of other factors people who play softball are aware of ... Real softball players would have to turn off their "real life" knowledge and experience. (Lesh & Lamon, 1992a, pp. 45–46)

This teacher's suggested improved version of the problem presents it in the form of the manager of the team selecting one of three players on the basis of their records to go in to bat with the game in a specific defined state. Lesh and Lamon (1992a, pp. 46–47) comment that "the main difference between the original softball problem and the modified softball problem is that the original asked a "school question" but did not provide any clues about the real-life issues or decisions that the response was intended to inform. By contrast, the modified problem ... calls for a realistic response to a situation that might really occur in the lives of the people who were asked to work on the problem". The revised problem still falls short of realism in the sense that, in a real situation, the manager would base a decision on much more complex aspects, such as the temperaments of the players, the assessed ability of the pitcher, and so on (cf the somewhat similar basketball problem discussed by Cooper (1992) (p. 151)).

Fifth, it is simply a matter of paying reasonable attention to weed out or fix carelessly composed questions that are unrealistic in that the numbers or conditions described would not occur in real life. For example, a little research will establish what a reasonable price would be for a given item (see the example cited at the head of Chapter 5). Questions involving unrealistic travelling at a constant speed can easily be fixed by rephrasing them in terms of approximations, or upper or lower bounds sensitive to the demands of the situation (which are usually, in any case, what people require). Unrealistically "easy" numbers should not be necessary with the availability of calculators, which don't care whether the numbers are "clean" or "dirty".

Swan (1993, p. 212) commented that the implementation of the laudable desire to introduce more applications into school examinations has been largely cosmetic, with questions merely "dressed up" in a realistic context. A particularly telling example he refers to shows a diagram of an ironing board, with various measurements given. The question is to work out the angle between the legs that support the ironing board, but as Swan points out, why would anyone ever want to know that? There is, he suggests, a "real" design question to be asked, namely, where should the "stops" under the top surface be placed so that the board could be adjusted for people of different heights?

Sixth, the belief that a word problem must have a single exact answer can be undermined by using problems calling for an appropriate mix of forms of answer. Answers that are acknowledged as estimations are the most obvious case, and these need to be tailored to the specific situation – how close an approximation is necessary, is a point or interval estimate appropriate, should one err above or below for safety, and so on. Again, discrimination training is indicated, i.e. presenting children with contrasting examples and discussing them (Barron et al., 1998).

Seventh, the belief that there is a single correct interpretation for any given situation can be counteracted by presenting examples where alternative situation models are plausible, and by having students who put forward different such models arguing their relative merits. A good example is the one reported by Lesh (personal communication; described in Chapter 7, pp. 135–136) about how scores awarded for quantity and quality for art projects should be amalgamated into a single score.

Puchalska and Semadeni (1987) point out that there is a traditional fear of exposing children to error of any sort and cite a counterargument by Brownell (1942, p. 440) who stated that:

> Part of real expertness in problem solving is the ability to differentiate between the reasonable and the absurd, the logical and the illogical. Instead of being "protected" from error, the child should many times be exposed to error and be encouraged to detect and to demonstrate what was wrong, and why.

Learning such skills and the associated critical disposition would help to produce children who could feel empowered to reject questions of the "How old is the captain?" type as nonsensical and become intellectually autonomous (Yackel & Cobb, 1996), developing into adults who would have some immunity against the use of mathematics as "propaganda" (Koblitz, 1981) or in an authoritarian and intimidatary manner (Schoenfeld, 1991).

Eighth, children could be encouraged to formulate problems arising from familiar situations in their own lives or stories in the media. Unfortunately, as Lave (1992, p. 77) has pointed out, children asked to make up problems about their everyday life do not make up problems about their experience, but rather problems that mimic the stereotyped word problems of school textbooks. This is certainly true of a large collection of such problems collected from children by Whimbey, Lochhead, and Potter (1990). Apart from being dressed up by references to cartoons, comic books, monsters, insects, and "sick" jokes that are supposed to make the problems interesting for children, they exhibit the flaws and limitations that we have criticized.

Seeing the world "through mathematical eyes" can be considered part of a mathematical disposition. Thus, "it is not sufficient that students acquire certain concepts, skills, and heuristics ... they should also get a feel for situations and opportunities to use these skills, and should become inclined to do so whenever appropriate" (De Corte et al., 1996, p. 508). Lave (1992, p. 87) comments:

> It is not clear just how much immersion in a mathematical culture is required, but at the end of this mathematical experience, its graduates should be able to infuse mathematical meaning into their everyday experience when they own the problem, and engage their intentions in mathematizing it.

The above recommendations constitute a major program for a thorough overhaul of word problems used in instruction (and assessment) to meet the criticisms detailed in Chapter 5. In the next section, we suggest how such a reform of word problems could be achieved within a modeling approach.

The modeling approach to word problems

In Chapter 7, we introduced the idea that word problems could usefully be reconceptualized as exercises in mathematical modeling, and here we elaborate on the nature of the modeling process, first in general terms, and then in relation to word problems. The advantages and difficulties associated with taking this perspective are then discussed.

The nature of the modeling process

A number of key characteristics of mathematical modeling may be identified.

First, the criterion for judging models is not absolute truth, but usefulness. A fundamental philosophical shift in perspective took place in modern times, whereby mathematical systems which previously were thought of as true descriptions of the real world came to be perceived rather as models, prime examples being Euclidean geometry and Newtonian mechanics. A model is useful if it makes predictions that are, to a required degree, accurate – within very wide ranges, Euclidean geometry and Newtonian mechanics work.

Second, the evaluation of the usefulness of models is relative to the goals of those developing and using them. Turing once said that "for certain purposes, a bowl of cold porridge may be a perfectly adequate model of the brain". The degree of precision that is appropriate will vary, depending on the purpose of the model. In considering the applicability of a mathematical model to a social phenomenon, human factors typically need to be taken into consideration. Communication of interpreted results of a modeling exercise may need to take account of particular target audiences.

Third, the choice of model is often influenced by the resources available. In the case of mathematical modeling these include the mathematical tools available to the modelers (or capable of being developed). For example, in school mechanics, a model for projectile motion that ignores all forces except gravity is often taught; it is within the mathematical competence of students at that stage (rarely, however, is the fact that it is a simplified model emphasized). Resources also include physical tools such as calculating devices. Increasingly so with the availability of massive computer power and numerical rather than analytical methods, mathematicians and others using mathematical modeling are less constrained than was previously the case. For example, in the early days of Factor Analysis, approximate but computationally tractable methods were used instead of more precise methods that required prohibitively complex calculation. Cost is also a factor – particularly in commercial applications, there is typically a trade-off between quick, cheap, and rough models as opposed to more elegant and precise models that are more expensive to develop and/or to apply.

Fourth, as a result of the above considerations and others, models always simplify (often drastically). Throughout this book, we have referred to the apparent willingness of children faced with word problems to ignore aspects of reality. However, children are not the only people that ignore reality – mathematicians and scientists also do so, by discussing frictionless planes, point masses and so on (Boote, 1998). The difference is that mathematicians and scientists do so mindfully, explicitly stating what simplifying assumptions are being made, and with a feel for the degree of inaccuracy that the simplification introduces relative to the goals of the exercise. A story related by Stewart (1987, p. 227) illustrates strikingly how mathematicians simplify in order to model. It concerns a farmer who, worried about his herd's milk production, commissioned an analysis from the local university. The research team was headed by a mathematician, and, when the farmer expectantly opened the 100-page report that resulted, he found that the first sentence was: "Consider a spherical cow".

Fifth, for a given phenomenon, there may be several competing alternative models. In physics, for some time, two major models for the nature of light co-existed: some phenomena were more readily explained in terms of a wave model, others in terms of a particle model. With the vast increase in modern times of models of social as well as physical phenomena (for example, in economics) there are many cases where there is no consensual model and proponents of rival models struggle for supremacy, often in a context of political ideologies.

Sixth, modeling is often an iterative process, whereby the evaluation of a current candidate model suggests some mismatch between the model and what is known about the phenomenon which, depending on how significant it is, triggers a revision of the model to take account of the discrepancy or a more fundamental rethink. A historical example is the Ptolemaic model for astronomy based on circular orbits, in which the basic model was progressively modified by the addition of epicycles to take account of deviations between observations and the predictions of the model. By contrast, the Galilean revolution represented a radical change to a model based on elliptical orbits.

Taking these considerations into account, we can elaborate the earlier schematic diagram of mathematical modeling (Figure 0.1, p. xii) to include major influences on the process of modeling (Figure 9.1).

The modeling perspective on word problems

Elaborating on the preliminary discussion in Chapter 7, the schematic representation of the modeling process shown in Figure 9.1 may be adapted to word problems.

Here the starting point of the process is reduced to a simplified description of a situation, presented primarily through text, though possibly supplemented by other forms of information such as drawings, data in various forms, and so

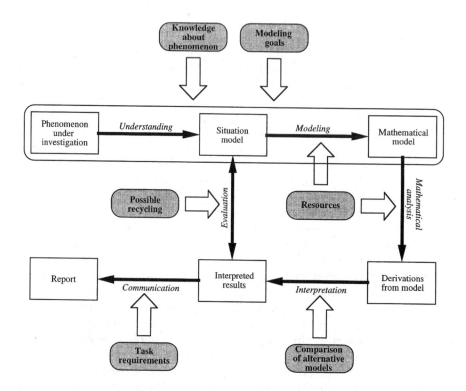

Fig. 9.1   An elaborated view of the modeling process (cf Fig. 0.1, p. xii).

on (see p. x). The first stage of modeling involves understanding the situation described and forming a situation model. Of course, this depends on having knowledge about the phenomena involved in the situation described. This is part of the first component for successful solving of application problems, as discussed in the introduction to this book (pp. xiv–xv). Having the knowledge is necessary but not sufficient – the knowledge must be accessed. As we have documented throughout this book, real-world knowledge is often suppressed, for reasons we have discussed at length.

The knowledge base does not have to be considered closed, as is typically the case in traditional word problem instruction. Students can extend it in the course of solving a problem by exploiting such resources as asking others, accessing sources of information, carrying out mental and physical experiments. In short, rather than work on such problems being seen as closed and isolated exercises in individual cognition, they can be relocated in a social/physical context in which cognition is distributed (Salomon, 1993a).

The next stage involves mathematizing, i.e. translating the situation model into mathematical form by identifying key quantities and relationships between them and expressing these by mathematical equations. (Although "mathematizing" is often used as a synonym for "mathematical modeling", we follow de Lange (1996, p. 68), in using the term more specifically for this translation part of the process). This second stage in getting from the problem presentation to a mathematical model obviously depends on another part of the knowledge base, namely knowledge about mathematical concepts, techniques, etc. Again, this part of the knowledge base need not be assumed to be closed. In contrast to the traditional assumption that computational techniques are learnt first, and then applied, problem solving can be used for conceptual development. This is a central principle of the Realistic Mathematics Education approach, a particularly clear example of this being the presentation of problems such as: "Santa Claus has his gifts distributed in the village by eight servants. Each has 23 parcels. How many parcels do they have altogether?" (Treffers, 1987, p. 200) and "Distribute 324 match stickers fairly among four children ... how many does each of them get?" (Treffers, 1987, p. 203) prior to students being taught any algorithm for column multiplication and division and followed by progressive schematization culminating in the formal algorithm. Accordingly, "during the entire process of learning column multiplication and division, all exercises are formulated by way of 'word problems', which aid rather than impede learning" (Freudenthal, 1991, p. 61).

A second component of successful problem solving is the use of heuristics. To the extent that word problems are true problems, in the sense that they involve constructing a solution that is not initially readily available to the solver, heuristics are implicated, as for problem solving in general. In the design experiment of Verschaffel, De Corte, Lasure, et al. (1999) reported in Chapter 6, a set of heuristics was explicitly taught within a general five-step problem-solving strategy (Table 6.5). This five-step strategy ((1) build a mental representation of the problem, (2) decide how to solve the problem, (3) execute the necessary calculations, (4) interpret the outcome and formulate an answer, (5) evaluate the solution) maps quite naturally onto the characterization of modeling shown in Figure 9.1 and described in Chapter 7 (pp. 130–131).

As indicated in Figure 9.1 and within the description of the nature of the modeling process given above, the production of a mathematical model may also be influenced by the goals implicit in the situation, imposed by the instructional context (e.g., a written assessment), or negotiated. For example, does the problem call for a precise or an approximate answer? To what extent should complications relating to realistic aspects of the situation be accommodated? In a real situation corresponding to that described in the problem, what would be the likely goals of the protagonist(s)? A clarification of the problem might portray a scenario in which the reasons for posing the problem are suggested. For example, the problem about the lift, heavily criticized by Cooper (1992) (see p. 69 and p. 150) noticeably lacks any indication of why anyone would want to know how many times the lift must go up to accommodate the 269 people in the morning rush.

Another component that has a bearing on the generation of a mathematical model is the nature of the resources available to the students. These resources include mathematical techniques known or potentially constructable in the course of working on the problem. Also of critical importance are mediating representational resources such as symbols, graphs and manipulatives. Increasingly, moreover, access to software modeling tools may be provided. For example, in Cobb's (1999) study of seventh grade students' explorations of distributions of data, two "minitools" were provided, such as a vertical bar that could be dragged along the axis of a graph showing a distribution to partition the data in various ways. Likewise, the SMART programs developed by the Cognition and Technology Group at Vanderbilt (Barron et al., 1998; Vye et al., 1997) included a "toolbox" of software for generating visual representations as problem-solving aids. As indicated in Figure 9.1 the availability or non-availability of these resources has an obvious effect at the stage at which mathematical analysis is applied to the mathematical model, but in anticipation of this effect it may influence the derivation of the mathematical model itself. Cobb (1999, p. 28) comments that "it should be apparent from the analysis of the sample episodes that the use of tools and symbols is integral to both the mathematical practices and the reasoning of the students who participate in them".

As shown schematically in Figure 7.2 (p. 136), there may not be a single undisputed way to model a given situation, as in the example about combining marks for quality and quantity, discussed in Chapter 7 (pp. 135–136). In Cobb's (1999) study, for example, children were asked to compare distributions of the lengths of time lasted for two brands of battery. The use of a minitool mediated several different models for reaching a decision (as opposed to the usual behavior in traditional mathematics classrooms, which would be to calculate the mean of every data set "regardless of the question at hand" (Cobb, 1999, p. 13). Although one brand of battery predominantly lasted longest, samples of the same brand also had the shortest life span. One student commented "I would rather have a consistent battery that I know will get me over 80 hours than one you just try to guess" (Cobb, 1999, pp. 15–16). As this example illustrates, debate about alternative models is naturally linked to goals of the respective modelers.

As emphasized at several points throughout the book, the baldly numerical results of manipulating the mathematical model need to be interpreted in relation to the situation model. As indicated in Figure 1.3 (p. 13), this stage is often omitted by children, as with the buses problem when children answer "31, remainder 12" or "31.33" as the number of buses required.

At this point, also, the interpreted results need to be evaluated in terms of the situation model. A child who has recognized that adding hot and cold water should result in a mixture that is luke-warm should realize at this stage than an answer of 120° F is not an appropriate answer for the addition of water at 80° F and 40° F. Depending on the perceived fit between the implications of the mathematical model and the situation model, it may be appropriate to revise or fine-tune the model to take account of minor or localized misfits, or to abandon the model and look for another, in the case of major incompatibility.

Finally, the task requirements for school word problems in current practice very rarely go beyond a bald reporting of the result of a calculation. This aspect of the activity can be enriched by asking students to provide arguments in support of a particular model, or to critically compare competing models. For example, the problem about combining quality and quantity scores described by Lesh and discussed in Chapter 7 (pp. 135–136) ends thus: "Your recommendation should include a justification to explain why your recommendation is better than other choices that could be made". As part of the research on the Jasper adventures by the Cognition and Technology Group at Vanderbilt, students in one of the studies had to report their solutions, and to discuss their solutions, on a local TV station (see p. 112).

In the introduction, we referred to the consensus that successful solving of application problems (as for mathematical problems in general) depends on the integration of four components, namely the knowledge base, exploitation of heuristics, metacognition, and affect. In relation to the characterization of word problem solving in terms of the modeling process presented above, the relevance of the first two has already been indicated. The component of metacognition that comprises skills and strategies for the regulation of problem solving is clearly important also at several stages, notably determining goals, building a situation model, making a solution plan, interpreting and evaluating the results. The model of competent problem solving used in the design experiment by Verschaffel, De Corte, and Lasure et al. (1999) (see pp. 96–107) may be seen as an externalization of the appropriate monitoring of the solution process.

Affect covers beliefs, attitudes and emotions. As extensively analyzed in Chapter 5 (pp. 58–71), it is clear that one of the strongest effects of standard forms of word problem instruction and practice is the set of beliefs that children acquire about the word problem game, such as that the mathematical computations required can be derived from superficial cues in the text, that any problem will require computational techniques that have recently been taught in class, that there is a single, exact answer to any problem, and so on.

Although most of the evidence is anecdotal, there is a clear consensus that children do not like word problems (Sowder, 1989). The popular view of word problems is summed up in a well-known cartoon by Gary Larson. With devil and flames in the background, a man is scanning bookshelves on which are arrayed hefty tomes with titles such as "Big Book of Story Problems", "Story Problems Galore" and so on. The succinct title is "Hell's Library".

Advantages and difficulties of the modeling perspective

There would be many advantages in adopting the modeling approach to word problems (Greer, 1993, 1997):

- Such an approach would undermine the achievement of apparent success through superficial strategies that is made possible by the stereotyped nature of word problems (Nesher, 1980; Reusser, 1988), whereby "most text-book problems are nothing more than poorly disguised exercises in one of the four basic exercises" (Gravemeijer, 1997, p. 390).

- Through such activities as questioning assumptions and debating the relative merits of rival models, students can be helped to become "flexible discourse and problem comprehenders" (Staub & Reusser, 1995, p. 302), and adaptive rather than routine experts (Hatano, 1988) at solving word problems (and see Wyndhamn & Säljö, 1997).

- The modeling perspective addresses the concerns raised by what Freudenthal (1991, p. 5) called "the poor permeability of the membrane separating classroom and school experience from life experience" (cf Resnick, 1987). As commented by Gravemeijer (1994, p. 100) "models are used as mediating tools to bridge the gap between situated knowledge and formal mathematics" – and, as argued by Nunes et al. (1993, p. 149) "between street and school mathematics".

- The modeling perspective is pervasive throughout mathematics (and science). Even what might be considered the starting point of mathematics, the natural numbers, can be interpreted as modeling multiple aspects of the world. Introducing students to this perspective early is arguably a key part of the process of nurturing a mathematical disposition, and it establishes continuity with more advanced forms of modeling in areas such as mechanics, data handling, and probability.

- An understanding of the modeling process and how mathematical models of social phenomena are constructed and interpreted, together with a critical stance towards such models, is an essential part of education for responsible citizenship (de Lange, 1996; Koblitz, 1981; Mukhopadhyay & Greer, in press; Schoenfeld, 1991). In *Descartes' Dream* Davis and Hersh (1986, p. xv) contend that "An emerging philosophy of applications and of computation must be concerned with the question: *why is mathematics used?*" and, in discussing this issue, declare:

We want to make a statement and draw a moral.

The statement, in brief is:

The social and physical worlds are being mathematized at an increasing rate.

The moral is:

We'd better watch it, because too much of it may not be good for us.

In the introduction (p. xi–xii), a number of pedagogical purposes for which word problems can be used was listed. If the modeling perspective is adopted, the following additions can be made to that list:

- To train students to think critically about the nature and uses of mathematical modeling (= as vehicles for developing modeling and critical skills).
- To develop new mathematical structures, notations, etc in the course of exploring the modeling of phenomena (= mathematical invention).
- To develop a propensity to see the world "through mathematical eyes" (= mathematical disposition).
- To provide a context for social interaction – discussion of the relative merits of rival models, communication of results etc (= socialization function).
- To deepen understanding of the nature of mathematics (= enculturation function).

While putting the modeling approach forward for serious consideration, it must be recognized that it not without major difficulties. For example, Hatano (1997, p. 385) questions whether the modeling approach is cost-effective:

It should be noted that the notion of mathematical modeling implies an inquiry that requires much time and effort, or in short, cost. Mathematical modeling is a kind of "understanding through comprehension activity", which is in contrast to "understanding through schema application" (Hatano, 1996).

... comprehension activity proceeds by deriving and testing predictions from each of the interpretations being considered, so it must be a time- and effort-consuming process. Scientific inquiry is often considered as prototypical of this activity, but it is surely welcome that educators encourage students to apply this activity to mathematics...

The question remains whether solving a typical wor(l)d problem through modeling activity brings about enough benefit to students to justify the high cost it requires to perform.

In response, we would contend that the modeling approach does not rule out "understanding through schema application". For example, the schema of direct proportionality is very powerful, and applicable to a wide variety of situations. The problem arises when such a schema is automatically triggered by superficial cues, i.e. students are given no training in discriminating between those cases where it is appropriate and those in which it is inappropriate or, in the extreme, nonsensical. Secondly, in terms of cost-benefit analysis, the cost of students being induced to indulge in violations of sense-making must be weighed on the other side of the balance. Do we want students to show merely "routine expertise" or should we aim for "adaptive expertise" (Hatano, 1988)?

A second difficulty that has been raised is the question of how much reference to the complexity of reality is appropriate. For example, Reusser and Stebler (1997a, p. 326) warn that:

> ... it might well be that students, if taking problems (simplistic word problems or more authentic ones) really seriously from a semantic or real-world point of view, might fail to model them by the use of mathematical means, not because of the weakness of their mathematical or real-world knowledge, but because of the virtual complexity of any real-world situation.

Likewise, Gravemeijer (1997, p. 392) refers to "an unsurmountable difference between solving authentic problems in reality and solving word problems in school mathematics" and points out that "In the classroom one is always dealing with reduction of reality. The question for the students, however, is which level of reduction is expected".

What is required is for students to "learn the delicate art of mathematizing – of taking complex situations, figuring out how to simplify them, and choosing the relevant mathematics to do the task" (Schoenfeld, 1987b, p. 37). At the same time, it is necessary to develop a sense of discrimination. Freudenthal (1991, p. 123) included as one of the "big strategies" for acquiring a mathematical attitude "identifying the mathematical structure within a context, if any is allowed, and barring mathematics where it does not apply".

As Cooper (1992, p. 235) commented in relation to the problem about the lift discussed earlier (p. 59, p. 150): "Some reference to the 'real' must be made, but not too much". Is encouraging children to use real world knowledge opening a Pandora's box? Some children interviewed by Selter (1994) who gave the answer 36 as the age of the captain on the boat with 26 sheep and 10 goats came up with explanations to justify this answer (see p. 62).

A particular problem concerns the degree of precision required in a model. One example was raised in discussion during a symposium with our research colleagues (Greer & Verschaffel, 1997). If you take real world considerations into account, can you really saw a piece of wood 2 meters long and get two pieces each 1 meter long (see Problem S2, Table 2.2), bearing in mind that the act of sawing removes a small amount of wood? Lewis Carroll in 1870 anticipated such questions with a puzzle:

> A stick I found that weighed two pound:
> I sawed it up one day
> In pieces eight of equal weight!
> How much did each piece weigh?

(Everybody says "a quarter of a pound" which is wrong.)

(Carroll, 1870, cited in Fisher, 1975, p. 64)

The solution offered is as follows (Fisher, p. 65):

In Shylock's bargain for the flesh was found
No mention of the blood that flowed around:
So when the stick was sawed in pieces eight,
The sawdust lost diminished from the weight.

Gravemeijer (1997, p. 393) commented as follows (emphasis added):

In the actual activity of sawing planks, one will in general disregard
the inaccuracy due to the serrated edge. Only in very special cases
would such a high precision be asked for. *In my view, this is an
important aspect of real-world problems. The students have to learn
to deliberate on estimation and rounding in relation to what precision the portrayed situation asks for.*

Thus, the degree of precision may be regarded, not as a difficulty, but as part of
what we want students to learn to make judgments about.

Such difficulties are most serious, we may suggest, when word problems are
presented in a context that precludes discussion, such as a student working alone
on a textbook problem, or sitting a written test. Within a context of discussion,
the degree of precision, the reasonableness of plausible assumptions, and so on,
may be negotiated. According to Yackel and Cobb (1996), it is a matter of inter-
actively constituting socio-mathematical norms. Such a perspective demands a
radical change in classroom culture from that which generally obtains at present.

Changing classroom culture

In this chapter, we have made substantial suggestions for design improvements
in word problems and a suggestion for a reconceptualized framework within
which to teach them. For such reforms to be effective, they must be embedded
in a classroom culture radically different from that which typically obtains and
is responsible, we have argued, for the detrimental effects documented in the
studies carried out by ourselves, our research collaborators, and others.

In Lave's (1992) terms, school mathematics is a particular form of practice
situated in a particular context. Van Oers (1998, p. 469) declared that:

The understanding of the relationship of human beings with their
immediate environment (or situation) is basic to all discussions about
human existence as are attempts to use these insights for educational purposes. The modern educational enterprise has basically developed into an institution for the design of contexts that are intended
for special purposes.

While, as Van Oers suggests (p. 470), "the redefinition of context opens up the possibility of designing environments for special purposes (leisure, education) that are based on new rules, materials, tools (including programming devices, scaffolds, curricula etc.), and special communicative strategies" it is also the case (p. 471) that "contexts for mathematics ... can emerge gradually – without explicit design – from a variety of socio-cultural determinants".

It seems clear that the context in which word problem practice is currently situated closely fits the latter description: current practice in the instruction of word problems shows no indication of design. Is it possible to move from such an emergent context for word problems to a designed context? Gravemeijer (1997, p. 392) cited the statement by Reusser and Stebler (1997a, p. 309) that mathematics, including word problems, in school is "inseparable from the (micro)cultural web of socio-cognitive practices of instruction, and of the materials used" and suggested that in order to change this, one has to change the didactical contract. Exemplary efforts to achieve this aim were presented in Chapter 6 and are discussed further below, along with other examples.

Changing the didactical contract

In Chapter 6, three design experiments were reported (Cognition and Technology Group at Vanderbilt, 1997; Verschaffel & De Corte, 1997b; Verschaffel, De Corte, Lasure, et al., 1999) that may be seen as exercises in changing the didactical contract in regard specifically to word problems, more broadly to the solving of application problems. Fundamental characteristics shared by all of these examples include:

- The use of more realistic, complex and challenging examples, in line with the detailed suggestions for the improvement of word problems made earlier in this chapter. In particular, the video-based presentation used in the Jasper studies allows a wealth of information to be presented, embedded in a coherent and interesting narrative, and so allowing very complex multi-step modeling tasks to be set.
- Explicit training in problem solving. In Verschaffel, De Corte, Lasure, et al. (1999), for example, a five-step problem solving strategy, incorporating eight specific heuristics, was used. The studies also exploit modeling by experts of strategic elements of the solution of application problems.
- Collaborative work in small groups and whole class discussions reflecting the social nature of activities implicit in the elaborated view of modeling described earlier in this chapter.
- Concerted efforts to create a culture within which students will develop appropriate beliefs about the nature of the modeling process. In Cobb's (1996) terms this means establishing socio-mathematical norms concerning, for example, what counts as a solution, what counts as an acceptable mathematical explanation, and so on. Evidence from the Jasper studies indicates changes in affect, with students becoming less anxious about mathematics, seeing it as more relevant to everyday life and more useful.

- Apart from the first study (Verschaffel & De Corte, 1997b), the teachers have been partners in the research, and considerable effort has been put into professional support, and into developing the teachers' own conceptions of solving application problems and of appropriate ways of teaching.

In general terms, all three examples may be characterized as developmental research (Cobb & Bowers, 1999; Gravemeijer, 1994) to the extent that they have been carried out in classrooms and have tested instructional sequence designed in reaction to research and theory. However, the first two are certainly limited by comparison with full-blown developmental research in that they were relatively short-term interventions and did not go through developmental cycles. The Jasper program of research, however, has developed over many years, and has gone through cycles whereby instructional design and classroom research have influenced each other. This approach to applying research to practice contrasts sharply with that characteristic of cognitive psychology which, according to Cobb and Bowers (1999, p. 12) "is seen to stand apart from and above instructional practice" and sees its role as "to generate an empirically substantiated body of principles that can serve as the primary resource to which teachers should turn for answers". The approach to word problems in Reed's (1999) recent book on the topic exemplifies this tradition.

Besides the didactical contract being changed, it can also be made more overt to the students through discussion, not just of specific models, but also of the nature of the modeling process per se. For instance, in the quotation at the head of this chapter, Keitel (1989) laments a lost opportunity for a teacher to discuss with students the idealization inherent in modeling relative to the complexity of reality. Explicit discussion of such topics as what constitutes a good problem, what counts as a good response, what counts as a good solution process (Verschaffel, De Corte, Lasure, et al., 1999) opens up the didactical contract to inspection instead of allowing it to remain implicit. Likewise, as has been discussed at various points (e.g., p. 175), it would seem reasonable for students to have a clearer idea, when taking realistic considerations into account, just how far they should go in this regard. The same remarks can be applied to making explicit the "assessment contract" (Van den Heuvel-Panhuizen, 1996, p. 102).

By way of one further illustration of the theme of being open with students about the nature of the activity in which they are engaged, it might be worth exploring the potential of humor for exploring aspects of word problems. The story quoted at the head of Chapter 7, for example, about the individual who considers that two doctors telling him he can drink two pints of beer every night sanctions his drinking four pints, makes, in a telling way, a fundamental point about the difference between $2 + 2 = 4$ as a statement of arithmetic and as a potential model for various situations. Likewise, Lewis Carroll, who has been quoted more than once, used humor to probe the relationship between language and mathematics. The popular children's book, *Math Curse* (Scieszka & Smith, 1995) has educational potential (possibly not envisaged by the authors) for provoking a discussion about the distinction between the surface structure and the deep structure of word problems, as in this series of questions:

> How many inches in a foot?
> How many feet in a yard?
> How many yards in a neighborhood?
> How many inches in a pint?
> How many feet in my shoes?

Individual and social perspectives as complementary
The elaborated characterization of the modeling process presented in Figure 9.1 and discussed above implies at several stages that modeling is inherently social in relation to:

* Knowledge about the situation being modeled which can be seen as open-ended and distributed as opposed to fixed and located in an individual mind.
* The goals of the models under consideration.
* The resources available to the modeler in terms of other people, sources of extra information, and cultural tools.
* Argumentation about the relative merits of alternative models.
* Evaluation of the interpreted results.
* Communication, which may need to be tailored to a specific audience.

Such a view of modeling contrasts totally with the predominant current practice of word problem solving whereby isolated students respond to informationally closed written problems. Although they were referring to students doing researchers' paper-and-pencil tests, the following statement by Wyndhamn and Säljö (1997, p. 380) applies also to students solving word problems in typical circumstances:

> The lack of a conversational partner responding to statements and the absence of a collective validating and scrutinising [of] statements, create a situation in which many productive elements may be lacking.

The contrast between the two situations is clearly demonstrated in Wyndhamn and Säljö's (1997) experiment (discussed on pp. 46–48) involving the problem about distances between B and C given the distances between A and B and A and C. Whereas on paper-and-pencil tests, children overwhelmingly give a single answer with no sign that they perceive anything possibly problematic, when the task is presented in the context of a reasonably natural discussion, the results are entirely different.

The above remarks apply in particular to standard methods of assessment (Wyndhamn & Säljö, 1997, p. 380):

> [An interesting question is] why there is a continued assumption that we learn about the knowledge and skills of people when we place

them in a communications situation which is so deviant as the individual test. In order to understand why this situation is considered to be the proper context in which a person's knowledge may be ascertained, an extensive analysis of the history of formal education, its role in society and the metaphorics of learning that have developed within institutionalized schooling would be necessary. In the dominant metaphorical construction of learning, knowledge is construed as something that individuals have as their private property and that they should be able to call upon in any context...

... In an alternative perspective, knowledge requires that objects and events can be constituted in relevant manners and by means of relevant discursive and practical tools.

Merely introducing discussion into the classroom is unlikely to have much effect, as the results obtained by Reusser and Stebler (1997a), summarized in Chapter 3 (pp. 35–38) illustrate. Despite interventions that included students working in pairs, and whole class discussions, the gains were modest. The research reported in Chapter 6, and discussed above, was based on the assumption that a more radical intervention is required to have strong and lasting results, namely the establishing of appropriate social and socio-mathematical norms in the classroom (Cobb, 1996). Examples of social norms cited by Cobb (1996, p. 88) include students being obliged to explain and justify their reasoning, and being obliged to listen to and understand others' explanations. As Cobb points out, these are not specific to mathematics, but equally applicable to any subject matter area, whereas socio-mathematical norms are specific to mathematics. A major aim of the design experiments described above was to develop a number of such norms in relation to the solving of application problems. The same aim was pursued in an even more systematic and explicit way in English's (1998) 12-week program for problem posing in middle-school classrooms. The first week is completely devoted to discussing and reflecting on problems, problem solving, and problem posing. These activities involve not only informal discussions on such topics as how student perceive mathematical problems, what they like and dislike about problems, whether they are good at solving them, but also having students argue their case in response to statements on debating cards such as "You learn more from working on a hard problem than from working on 10 easy problems", "Working problems on your own is better than working with a partner or in a group", or "You can learn more by making up your own problems than you can by working with the problems in the textbook".

The emphasis on social processes is in line with a major shift in mathematics educational research. Nevertheless, there is a danger of imbalance, and of forgetting the continuing contribution of psychological analyses of individual cognitive behavior (Greer, 1996). Avoiding such imbalance, Cobb (1996) works within an interactionist framework that involves the coordination of social and psychological perspectives (Figure 9.2) whereby "individual students'

Social perspective	Psychological perspective
Classroom social norms	Beliefs about own role, others' roles and the general nature of mathematical activity in school
Sociomathematical norms	Mathematical beliefs and values
Classroom mathematical practices	Mathematical conceptions

Fig. 9.2   An interpretative framework for analyzing mathematical activity in social context (Cobb, 1996, p. 87). Copyright 1996 by the author. Reprinted with permission.

mathematical interpretations, solutions, explanations, and justifications are seen not only as individual acts, but simultaneously as acts of participating in collective or communal classroom processes" (p. 85).

Similarly, Salomon and Perkins (1998) discuss the general relationship between social and individual learning, and Salomon (1993b, p. 111), in a chapter entitled "No distribution without individual's cognition: a dynamic interactional view" comment that: "As is the case with so many other newly coined terms, "distributed cognitions" strongly illuminates one facet of an issue, sending to dark oblivion others" (and see also Inagaki, Hatano, & Morita, 1998). In this respect, we should acknowledge that a limitation of the work reported in this book is that we have paid little attention to the detailed analysis of the psychological processes in individual students – redressing this imbalance would be an appropriate focus for future research.

Counting the cost

In this chapter we have laid out radical suggestions for changing the role of, and ways of teaching, word problems within the mathematics curriculum. The cost of doing so would be very great, in terms of the human resources that would be required. In particular, like any reform of mathematics education, it depends on teachers, and the evidence we have presented indicates that fundamental changes in the conceptualization of word problems on the part of most teachers would be needed. The examples of design experiments discussed in Chapter 6 and earlier in this chapter are promising in the sense that they show how enriched forms of teaching can produce positive effects. The problem is to reproduce these effects at a systemic level, implying a coherent and sustained campaign that integrates all components of the instructional environment, in particular, professional development of teachers, textbooks, and assessment (De Corte, in press).

As Jacob (1997) has pointed out, political and institutional factors wider than schools are at work. She cites, for example, the conclusion of Cobb, Wood, and Yackel (1993) that the school district's required objectives for mathematics, and state-mandated accountability tests of "basic mathematics" reinforced traditional instructional practices. De Corte et al. (1996, p. 534) suggested that: "Perhaps the enlightenment of political decision-makers, and other groups such as parents, administrators, and the public in general ... is the biggest educational challenge facing reformers". In short, a systemic implementation of the approach that we have advocated could only be achieved with immense difficulty and at great cost. But what of the cost of maintaining the status quo?

However it may be rationalized, we instinctively find it alarming that children are prepared to give the age of the captain. Less dramatically, but perhaps more seriously, we are concerned about children who answer word problems on the basis of superficial cues, going from text to solution without passing through a thinking brain. The evidence laid out in the early chapters of this book affronts our conception of how children should come to regard mathematics, and we are not alone in this reaction. Freudenthal (1991, p. 70) commented:

> In the textbook context each problem has one and only one solution:
> There is no access for reality, with its unsolvable and multiply
> solvable problems. The pupil is supposed to discover the pseudo-
> isomorphisms envisaged by the textbook author and to solve prob-
> lems, which look as though they were tied to reality, by means of
> these pseudo-isomorphisms. Wouldn't it be worthwhile investigating
> whether and how this didactic breeds an anti-mathematical attitude
> and why the children's immunity against this mental deformation is
> so varied?

It is arguable that the complex of practices in which word problems is embedded provides a prototypical example of what is wrong with mathematics education in general. The cost of continuing to teach mathematics, and word problems in particular, according to current practices may be a population in which the majority of people remain alienated from mathematics.

References

Anand, P., & Ross, S. M. (1987). Using computer-assisted instruction to personalize math learning materials for elementary school children. *Journal of Educational Psychology, 79*, 72–79.

Ascher, M., & Ascher, R. (1997). Ethnomathematics. In A. B. Powell & M. Frankenstein (Eds.), *Ethnomathematics: Challenging Eurocentrism in mathematics education* (pp. 25–50). New York: State University of New York Press.

Austin, J. L. (1975). *How to do things with words*. Oxford: Clarendon.

Australian Education Council (1990). *A national statement on mathematics for Australian schools*. Carlton, Australia: Curriculum Cooperation.

Barron, B., Schwartz, D. L., Vye, N. J., Moore, A., Petrosino, A., Zech, L., Bransford, J. D., & The Cognition and Technology Group at Vanderbilt (1998). Doing with understanding: Lessons from research on problem-based and project-based learning. *The Journal of the Learning Sciences, 7*, 271–311.

Baruk, S. (1985). *L'âge du capitaine. De l'erreur en mathématiques*. Paris: Seuil.

Becker, J. R. (1995). Women's ways of knowing in mathematics. In P. Rogers & G. Kaiser (Eds.) *Equity in mathematics education: Influences of feminism and culture* (pp. 163–174). London: Falmer Press.

Beckers, D. (1999). *"My little mathematicians!" Paedagogic ideals in Dutch mathematics education 1790–1850*. (Internal report.) Nijmegen, The Netherlands: University of Nijmegen, Department of Mathematics.

Belensky, M. F., Clinchy, B. M., Goldberger, N. R., & Tarule, J. M. (1985). *Women's ways of knowing: The development of self, voice, and mind*. New York: Basic Books.

Bell, A., Burkhardt, H., & Swan, M. (1992a). Balanced assessment of mathematical performance. In R. Lesh & S. Lamon (Eds.), *Assessment of authentic performance in school mathematics* (pp. 119–144). Washington, DC: American Association for the Advancement of Science.

Bell, A., Burkhardt, H., & Swan, M. (1992b). Moving the system: The contributions of assessment. In R. Lesh & S. Lamon (Eds.), *Assessment of authentic performance in school mathematics* (pp. 177–193). Washington, DC: American Association for the Advancement of Science.

Bender, P. (1985). Der Primat der Sache in Sachrechnen. *Sachunterricht in Mathematik in der Primarstufe, 13,* 141–147.

Bloor, D. (1994). What can the sociologist of knowledge say about 2 + 2 = 4? In P. Ernest (Ed.), *Mathematics, education and philosophy: An international perspective* (pp. 21–32). London: Falmer Press.

Boaler, J. (1994). When do girls prefer football to fashion? An analysis of female under-achievement in relation to "realistic" mathematical contexts. *British Educational Research Journal, 20,* 551–564.

Boaler, J. (1997a). Equity, empowerment and different ways of knowing. *Mathematics Education Research Journal, 9,* 325–342.

Boaler, J. (1997b). *Experiencing school mathematics: Teaching styles, sex and setting.* Buckingham, England: Open University Press.

Boote, D. (1998). Physics word problems as exemplars for enculturation. *For the Learning of Mathematics, 18 (2),* 28–33.

Bransford, J. D., & Stein, B. S. (1993). *The IDEAL problem solver* (2nd ed.). New York: Freeman.

Brissiaud, R. (1988). De l'âge du capitaine à l'âge du berger. Quel contrôle de la validité d'un énoncé de problème au CE2? *Revue Française de Pédagogie, 82,* 23–31.

Brousseau, G. (1980). L'Èchec et le contrat. *Recherches, 41,* 177–182.

Brousseau, G. (1984). The crucial role of the didactical contract in the analysis and con-struction of situations in teaching and learning mathematics. In H. G. Steiner (Ed.), *Theory of mathematics education* (Occasional Paper 54, pp. 110–119). Bielefeld, Germany: Institut für Didaktik der Mathematik, Universität Bielefeld.

Brousseau, G. (1990). Le contrat didactique: Le milieu. *Recherches en Didactique de Mathématiques, 9,* 308–336.

Brousseau, G. (1997). *Theory of didactical situations in mathematics* (Edited and trans-lated by N. Balacheff, M. Cooper, R. Sutherland, & V. Warfield). Dordrecht, The Netherlands: Kluwer.

Brown, J. S., Collins, A., & Duguid, P. (1989). Situated cognition and the culture of learning. *Educational Researcher, 18 (1),* 32–42.

Brown, M. (1993). Assessment in mathematics education: Developments in philosophy and practice in the United Kingdom. In M. Niss (Ed.), *Cases of assessment in math-ematics education* (pp. 71–84). Dordrecht, The Netherlands: Kluwer.

Brown, S. I., & Walter, M. I. (Eds.). (1993). *Problem posing: Reflections and applica-tions.* Hillsdale, NJ: Lawrence Erlbaum Associates.

Brownell, W. A. (1942). Problem solving. In N. B. Henry (Ed.), *The psychology of learn-ing (41st Yearbook of the National Society for the Study of Education, Part 2)* (pp. 400–415). Bloomington, IL: Public School Publishing Co.

Burkhardt, H. (1981). *The real world and mathematics.* Glasgow, UK: Blackie & Son.

Burkhardt, H. (1994). Mathematical applications in school curriculum. In T. Husén & T. N. Postlethwaite (Eds.), *The international encyclopedia of education* (2nd ed.) (pp. 3621–3624). Oxford/New York: Pergamon Press.

Cai, J., & Silver, E. A. (1995). Solution processes and interpretations of solutions in solving division-with-remainder story problems: Do Chinese and U. S. students have similar difficulties? *Journal for Research in Mathematics Education, 26,* 491–497.

Caldwell, J. H., & Goldin, G. A. (1979). Variables affecting word problem difficulty in elementary school mathematics. *Journal for Research in Mathematics Education, 10,* 323–336.

Caldwell, L. (1995). *Contextual considerations in the solution of children's multiplication and division word problems.* Unpublished undergraduate thesis, Queen's University, Belfast, Northern Ireland.

Calvino, I. (1988). *Six memos for the next millennium.* Cambridge, MA: Harvard University Press.

Carpenter, T. P., Moser, J. M., & Romberg, T. A. (Eds.). (1982). *Addition and subtraction: A cognitive perspective.* Hillsdale, NJ: Lawrence Erlbaum Associates.

Carpenter, T. P., & Fennema, E. (1992). Cognitively guided instruction: Building on the knowledge of students and teachers. *International Journal of Educational Research, 17,* 457–470.

Carpenter, T. P., Lindquist, M. M., Matthews, W., & Silver, E. A. (1983). Results of the third NAEP mathematics assessment: Secondary school. *Mathematics Teacher, 76,* 652–659.

Carraher, T. N., Carraher, D. W., & Schliemann, A. D. (1985). Mathematics in streets and schools. *British Journal of Developmental Psychology, 3,* 21–29.

Chevallard, Y. (1988). *Sur l'analyse didactique. Deux études sur les notions de contrat et de situation* (Publications de l'IREM d'Aix-Marseille). Aix-en-Provence/Marseille, France: Institut de Recherche sur l'Enseignement des Mathématiques.

Clarke, D. (1996). Assessment. In A. J. Bishop, K. Clements, C. Keitel, J. Kilpatrick, & C. Laborde (Eds.), *International handbook of mathematics education, Vol. 1,* pp. 327–370. Dordrecht, The Netherlands: Kluwer.

Clarke, D., & Stephens, W. M. (1996). The Ripple Effect: The instructional impact of the systemic introduction of performance assessment in mathematics. In M. Birembaum & F. Dochy (Eds.), *Alternatives in assessment of achievements, learning processes and prior knowledge* (pp. 63–92). Dordrecht, The Netherlands: Kluwer.

Cobb, P. (1995). Cultural tools and mathematical learning: A case study. *Journal for Research in Mathematics Education, 26,* 362–385.

Cobb, P. (1996). Accounting for mathematical learning in the social context of the classroom. In C. Alsina, J. M. Alvarez, B. Hodgson, C. Laborde, & A. Pérez (Eds.), *Eighth International Congress on Mathematical Education: Selected Lectures* (pp. 85–99). Seville, Spain: S. A. E. M. Thales.

Cobb, P. (1999). Individual and collective mathematical development: The case of statistical data analysis. *Mathematical Thinking and Learning, 1,* 5–43.

Cobb, P., & Bowers, J. (1999). Cognitive and situated learning: Perspectives in theory and practice. *Educational Researcher, 28 (2),* 4–15.

Cobb, P., Wood, T., & Yackel, E. (1993). Discourse, mathematical thinking, and classroom practice. In E. Forman, N. Minick, & A. Stone (Eds.), *Contexts for learning: Social cultural dynamics in children's development* (pp. 91–120). New York: Springer.

Cobb, P., Yackel, E., & Wood, T. (1992). A constructivist alternative to the representational view of mind in mathematics education. *Journal for Research in Mathematics Education, 23,* 2–33.

Cockcroft, W. H. (1982). *Mathematics counts* (Report of the Committee of Inquiry into the Teaching of Mathematics in Schools). London: Her Majesty's Stationery Office.

Cognition and Technology Group at Vanderbilt (1997). *The Jasper project: Lessons in curriculum, instruction, assessment, and professional development.* Mahwah, NJ: Lawrence Erlbaum Associates.

Cohen, J. (1988). *Statistical power analysis for the behavioral sciences.* Hillsdale, NJ: Lawrence Erlbaum Associates.

Colebrooke (1967). *Translation of Lilavati by Bhaskaracharyya.* Allahabad, India: Kitab Mahal.

Cooper, B. (1992). Testing National Curriculum mathematics: Some critical comments on the treatment of "real" contexts in mathematics. *The Curriculum Journal, 3,* 231–243.

Cooper, B. (1994). Authentic testing in mathematics? The boundary between everyday and mathematical knowledge in National Curriculum testing in English schools. *Assessment in Education, 1,* 143–166.

Cooper, B., & Dunne, M. (1998a). Anyone for tennis? Social class differences in children's responses to National Curriculum mathematics testing. *Sociological Review, 46,* 115–148.

Cooper, B., & Dunne, M. (1998b, September). *Social class, gender, equity and National Curriculum tests in mathematics.* Paper presented at Conference on Mathematics Education and Society, Nottingham, England.

Cooper, B., Dunne, M., & Rodgers, N. (1997, April). *Social class, gender, item type and performance in national tests of primary school mathematics: Some research evidence from England.* Paper presented at the Annual Meeting of the American Educational Research Association, Chicago.

Cramer, K., Post, T., & Currier, S. (1993). Learning and teaching ratio and proportion: Research implications. In D. T. Owens (Ed.), *Research ideas for the classroom: Middle grades mathematics* (pp. 159–178). New York: Macmillan.

D'Ambrosio, U. (1999). Literacy, matheracy, and technoracy: A trivium for today. *Mathematical Thinking and Learning, 1,* 131–153.

Davis, P. J., & Hersh, R. (1981). *The mathematical experience.* Boston: Birkhauser.

Davis, P. J., & Hersh, R. (1986). *Descartes' dream: The world according to mathematics.* Sussex, England: Harvester.

Davis, R. B. (1989). The culture of mathematics and the culture of schools. *Journal of Mathematical Behavior, 8,* 143–160.

Davis-Dorsey, J. K., Ross, S. M., & Morrison, G. R. (1991). The role of rewording and context personalization in the solving of mathematical word problems. *Journal of Educational Psychology, 83,* 61–68.

Damarin, S. K. (1995). Gender and mathematics from a feminist standpoint. In W. G. Secada, E. Fennema, & L. B. Adajian (Eds.), *New directions for equity in mathematics education* (pp. 242–257). New York: Cambridge University Press.

De Bock, D., Verschaffel, L., & Janssens, D. (1998). The influence of cognitive and visual scaffolds on the predominance of the linear model. In A. Olivier & K. Newstead (Eds.), *Proceedings of the 22nd Conference of the International Group for the Psychology of Mathematics Education, Vol. 2,* pp. 240–247. Stellenbosch, South Africa: University of Stellenbosch.

De Bock, D., Verschaffel, L., Janssens, D., & Rommelaere, R. (1999). What causes improper proportional reasoning: The problem or the problem formulation? In O. Zaslavsky (Ed.), *Proceedings of the 23rd Conference of the International Group for the Psychology of Mathematics Education, Haifa, Israel, July 25–30 1999. Vol. 2* (pp.

241–248). Haifa, Israel: Technion B Israel Institute of Technology, Department of Education in Technology and Science.

De Corte, E. (1995). Fostering cognitive growth: A perspective from research on mathematics learning and instruction. *Educational Psychologist, 30,* 37–46.

De Corte, E. (in press). Marrying theory building and the improvement of school practice: A permanent concern for instructional psychology. *Learning and Instruction, 10.*

De Corte, E., Greer, B., & Verschaffel, L. (1996). Learning and teaching mathematics. In D. Berliner & R. Calfee (Eds.), *Handbook of educational psychology* (pp. 491–549). New York: Macmillan.

De Corte, E., & Verschaffel, L. (1985). Beginning first graders' initial representation of arithmetic word problems. *Journal of Mathematical Behavior, 4,* 3–21.

De Corte, E., & Verschaffel, L. (1989). Teaching word problems in the primary school: What research has to say to the teacher. In B. Greer & G. Mulhern (Eds.), *New developments in teaching mathematics* (pp. 85–106). London: Routledge.

De Corte, E., Verschaffel, L., Janssens, V., & Joillet, L. (1985). Teaching word problems in the first grade: A confrontation of educational practice with results of recent research. In T. A. Romberg (Ed.), *Using research in the professional life of mathematics teachers* (pp. 186–195). Madison, WI: Center for Educational Research, University of Wisconsin.

DeFranco, T. C., & Curcio, F. R. (1997). A division problem with a remainder embedded across two contexts: Children's solutions in restrictive versus real-world settings. *Focus on Learning Problems in Mathematics, 19 (2),* 58–72.

Dekker, R. (1996). Use your common sense. In C. Keitel, U. Gellert, E. Jablonka, & M. Müller (Eds.), *Mathematics (education) and common sense* (pp. 39–46). Berlin: Freie Universität Berlin.

de Lange, J. (1993). Innovation in mathematics education using applications: Progress and problems. In J. de Lange, C. Keitel, I. Huntley, & M. Niss (Eds.), *Innovation in mathematics education by modelling and applications* (pp. 3–19). Chichester, England: Ellis Horwood.

de Lange, J. (1995). Assessment: No change without problems. In T. A. Romberg (Ed.), *Reform in school mathematics and authentic assessment* (pp. 87–172). Albany, NY: State University of New York Press.

de Lange, J. (1996). Using and applying mathematics in education. In A. J. Bishop, K. Clements, C. Keitel, J. Kilpatrick, & C. Laborde (Eds.), *International handbook of mathematics education, Vol. 1,* pp. 49–97. Dordrecht, The Netherlands: Kluwer.

de Lange, J. (1998). Real problems with real world mathematics. In C. Alsina, J. Alvarez, M. Niss, A. Pérez, L. Rico, & A. Sfard (Eds.), *Proceedings of the Eighth International Congress on Mathematical Education* (pp. 83–110). Seville, Spain: S. A. E. M. Thales.

Dowling, P. C. (1991). Gender, class and subjectivity in mathematics: A critique of Humpty Dumpty. *For the Learning of Mathematics, 11 (1),* 2–8.

Dowling, P. C. (1996). A sociological analysis of school mathematics texts. *Educational Studies in Mathematics, 31,* 389–415.

Dowling, P. C. (1998). *The sociology of mathematics education: Mathematical myths/ pedagogic texts.* London: Falmer.

Eco, U. (1994). *Six walks in the fictional woods.* Cambridge, MA: Harvard University Press.

Ellerton, N. F. (1989). The interface between mathematics and language. *Australian Journal of Reading, 12,* 92–102.

English, L. D. (1998, April). *Problem posing in middle-school classrooms*. Paper presented at the Annual Conference of the American Educational Research Association, San Diego, CA.

Fennema, E., & Loef, M. (1992). Teachers' knowledge and its impact. In D. A. Grouws (Ed.), *Handbook of research on learning and teaching mathematics* (pp. 147–164). New York: Macmillan.

Fischbein, E. (1987). *Intuition in science and mathematics: An educational approach*. Dordrecht, The Netherlands: Reidel.

Fischbein, E., Deri, M., Nello, M. S., & Marino, M. S. (1985). The role of implicit models in solving verbal problems in multiplication and division. *Journal of Research in Mathematics Education, 16,* 3–17.

Fisher, J. (Ed.). (1975). *The magic of Lewis Carroll*. London: Penguin.

Frankenstein, M. (1989). *Relearning mathematics: A different third R - radical math(s)*. London: Free Association Books.

Frankenstein, M. (1996). *Critical mathematical literacy: Teaching through real real-life math word problems*. In T. Kjaergard, A. Kvamme, & N. Linden (Eds.), *Numeracy, race, gender, and class. Proceedings of the 3rd International Conference on Political Dimensions of Mathematics Education* (pp. 59–76). Landas, Norway: Casper Vorlag.

Frederiksen, J. R., & Collins, A. (1989). A systems approach to educational testing. *Educational Researcher, 18 (9),* 27–32.

Freudenthal, H. (1982). Fiabilité, validité et pertinence – critères de la recherche sur l'enseignement de la mathématique. *Educational Studies in Mathematics, 13,* 395–408.

Freudenthal, H. (1991). *Revisiting mathematics education*. Dordrecht, The Netherlands: Kluwer.

Fuson, K. C. (1992). Research on whole number addition and subtraction. In D. A. Grouws (Ed.), *Handbook of research on mathematics teaching and learning* (pp. 243–275). New York: Macmillan.

Gerofsky, S. (1996). A linguistic and narrative view of word problems in mathematics education. *For the Learning of Mathematics, 16 (2),* 36–45.

Goldin, G. A., & McClintock, E. (Eds.). (1984). *Task variables in mathematical problem solving*. Philadelphia: Franklin.

Goodell, J. E., & Parker, L. H. (in press). Creating a connected, equitable mathematics classroom. In W. Atweh, H. Forgasz, & B. Nebres (Eds.), *Socio-cultural aspects in mathematics education*. Mahwah, NJ: Lawrence Erlbaum Associates.

Gravemeijer, K. (1994). *Developing realistic mathematics education*. Utrecht, The Netherlands: Freudenthal Institute, University of Utrecht.

Gravemeijer, K. (1997). Solving word problems: A case of modelling? *Learning and Instruction, 7,* 389–397.

Greeno, J. G., & Middle School Mathematics Through Applications Project Group (1998). The situativity of knowing, learning, and research. *American Psychologist, 53,* 5–26.

Greer, B. (1987). Understanding of arithmetical operations as models of situations. In J. Sloboda & D. Rogers (Eds.), *Cognitive processes in mathematics* (pp. 60–80). Oxford: Oxford University Press.

Greer, B. (1992). Multiplication and division as models of situations. In D. A. Grouws (Ed.), *Handbook of research on mathematics teaching and learning* (pp. 276–295). New York: Macmillan.

Greer, B. (1993). The modeling perspective on wor(l)d problems. *Journal of Mathematical Behavior, 12,* 239–250.

Greer, B. (1996). Theories of mathematics education: The role of cognitive analyses. In L. P. Steffe, P. Nesher, P. Cobb, G. A. Goldin, & B. Greer (Eds.), *Theories of mathematical learning* (pp. 179–196). Mahwah, NJ: Lawrence Erlbaum Associates.

Greer, B. (1997). Modelling reality in mathematics classrooms: The case of word problems. *Learning and Instruction, 7,* 293–307.

Greer, B., & Verschaffel, L. (Eds.). (1997). Modelling reality in mathematics classrooms (Special issue). *Learning and Instruction, 7,* 293–397.

Harel, G., & Confrey, J. (1994). *The development of multiplicative reasoning in the learning of mathematics.* Albany, NY: State University of New York Press.

Hatano, G. (1988). Social and motivational bases for mathematical understanding. *New Directions for Child Development, 41,* 55–70.

Hatano, G. (1996). A conception of knowledge acquisition and its implications for mathematics education. In L. P. Steffe, P. Nesher, P. Cobb, G. A. Goldin, & B. Greer (Eds.), *Theories of mathematical learning* (pp. 197–217). Mahwah, NJ: Lawrence Erlbaum Associates.

Hatano, G. (1997). Commentary: Cost and benefit of modelling activity. *Learning and Instruction, 7,* 383–387.

Herbst, P., & Kilpatrick, J. (1999). Pour lire Brousseau. *For the Learning of Mathematics, 19 (1),* 3–10.

Hersh, R. (1997). *What is mathematics, really?* New York: Oxford University Press.

Hidalgo, M. C. (1997). *L'activation des connaissances à propos du monde réel dans la résolution de problèmes verbaux en arithmétique.* Unpublished doctoral dissertation, Université Laval, Québec, Canada.

Hiebert, J., & Behr, M. (1989). *Number concepts and operations in the middle grades.* Reston, VA: National Council of Teachers of Mathematics.

Houston, S. K., Blum, W., Huntley, I., & Neill, N. T. (Eds.). (1997). *Teaching and learning mathematical modelling.* Chichester, England: Albion Publishing.

Inagaki, K., Hatano, G., & Morita, E. (1998). Construction of mathematical knowledge through whole-class discussion. *Learning and Instruction, 8,* 503–526.

Institut de Recherche sur l'Enseignement des Mathématiques (IREM) de Grenoble (1980). *Bulletin de l' Association des professeurs de Mathématique de l' Enseignement Public, no 323,* 235–243.

Jacob, E. (1997). Context and cognition: Implications for educational innovators and anthropologists. *Anthropology and Education Quarterly, 28 (1),* 3–21.

Johnson, M. (1987). *The body in the mind.* Chicago: University of Chicago Press.

Joshua, S., & Dupin, J. J. (1993). *Introduction à la didactique des sciences et des mathématiques.* Paris: Presses Universitaires de France.

Keitel, C. (1989). Mathematics education and technology. *For the Learning of Mathematics, 9 (1),* 7–13.

Keitel, C. (1993). Implicit mathematical models in social practice and explicit mathematics teaching by applications. In J. de Lange, C. Keitel, I. Huntley, & M. Niss (Eds.), *Innovation in mathematics education by modelling and applications* (pp. 19–31). Chichester, England: Ellis Horwood.

Keitel, C. (1996). Discussion paper: Mathematics (education) and common sense. In C. Keitel, U. Gellert, E. Jablonka, & M. Müller (Eds.), *Mathematics (education) and common sense* (pp. 14–20). Berlin: Freie Universität Berlin.

Kilpatrick, J. (1985). A retrospective account of the past twenty-five years of research on teaching mathematical problem-solving. In E. A. Silver (Ed.), *Teaching and learning mathematical problem-solving: Multiple research perspectives* (pp. 1–16). Hillsdale, NJ: Lawrence Erlbaum Associates.

Kilpatrick, J. (1987). Problem formulating: Where do good problems come from? In A. H. Schoenfeld (Ed.), *Cognitive science and mathematics education* (pp. 123–147). Hillsdale, NJ: Lawrence Erlbaum Associates.

Kintsch, W., & Greeno, J. G. (1985). Understanding and solving arithmetic word problems. *Psychological Review, 92,* 109–129.

Kirschner, D., & Whitson, J. A. (Eds.) (1997). *Situated cognition theory: Social, semiotic, and psychological perspective.* Mahwah, NJ, Lawrence Erlbaum Associates.

Koblitz, N. (1981). Mathematics as propaganda. In L. A. Steen (Ed.), *Mathematics tomorrow* (pp. 111–120). New York: Springer.

Krutetskii, V. A. (1976). *The psychology of mathematical abilities in school children.* (J. Teller, Trans., J. Kilpatrick, & J. Wirszup, Eds.) Chicago, University of Chicago Press.

Lajoie, S. (1995). A framework for authentic assessment in mathematics. In T. A. Romberg (Ed.), *Reform in school mathematics and authentic assessment* (pp. 19–37). Albany, NY: State University of New York Press.

Lakoff, G., & Nunez, R. E. (1997). The metaphorical structure of mathematics: Sketching out cognitive foundations for a mind-based mathematics. In L. D. English (Ed.), *Mathematical reasoning: Analogies, metaphors, and images* (pp. 21–89). Mahwah, NJ: Lawrence Erlbaum Associates.

Lave, J. (1988). *Cognition in practice: Mind, mathematics and culture in everyday life.* Cambridge: Cambridge University Press.

Lave, J. (1992). Word problems: A microcosm of theories of learning. In P. Light & G. Butterworth (Eds.), *Context and cognition: Ways of learning and knowing* (pp. 74–92). New York: Harvester Wheatsheaf.

Lesh, R., & Lamon, S. (1992a). Assessing authentic mathematical performance. In R. Lesh & S. Lamon (Eds.), *Assessment of authentic performance in school mathematics* (pp. 17–62). Washington, DC: American Association for the Advancement of Science.

Lesh, R., & Lamon, S. (1992b). *Assessment of authentic performance in school mathematics.* Washington, DC: American Association for the Advancement of Science.

Lester, F., Garofalo, J., & Kroll, D. (1989). *The role of metacognition in mathematical problem solving. A study of two grade seven classes.* Final report to the National Science Foundation of NSF Project MDR 85–50346.

Libbrecht, U. (1973). *Chinese mathematics in the thirteenth century: The Shu-shu chiu-chang of Ch'in Chiu-shao.* Cambridge, MA: MIT Press.

Luria, A. R. (1976). *Cognitive development. Its cultural and social foundations.* Cambridge, MA: Harvard University Press.

Markovitz, Z., Hershkowitz, R., & Bruckheimer, M. (1984). Algorithm leading to absurdity, leading to conflict, leading to algorithm review. In B. Southwell, R. Eyland, M. Cooper, J. Confrey, & K. Collis (Eds.), *Proceedings of the Eighth International Conference for Psychology of Mathematics Education* (pp. 244–250). Sydney, Australia: International Group for the Psychology of Mathematics Education.

Menon, R. (1993, April). *The role of context in student-constructed questions.* Paper presented at the Annual Conference of the American Educational Research Association, Atlanta, GA.

Ministerie van de Vlaamse Gemeenschap (1997). *Gewoon basisonderwijs: ontwikkelingsdoelen en eindtermen. Besluit van mei '97 en decreet van juli '97* [Educational standards for the elementary school]. Brussels, Belgium: Departement Onderwijs, Centrum voor Informatie en Documentatie.

Mukhopadhyay, S., & Greer, B. (in press). Modelling with purpose: Mathematics as a critical tool. In B. Atweh, H. Forgasz, & B. Nebres (Eds.), *Socio-cultural aspects in mathematics education.* Mahwah, NJ: Lawrence Erlbaum Associates.

National Council of Teachers of Mathematics (1989). *Curriculum and evaluation standards for school mathematics.* Reston, VA: National Council of Teachers of Mathematics.

National Council of Teachers of Mathematics (1995). *Assessment standards for school mathematics.* Reston, VA: National Council of Teachers of Mathematics.

Nesher, P. (1980). The stereotyped nature of school word problems. *For the Learning of Mathematics, 1 (1),* 41–48.

Niss, M. (1993a). *Cases of assessment in mathematics education.* Dordrecht, The Netherlands: Kluwer.

Niss, M. (1993b). *Investigations into assessment in mathematics education.* Dordrecht, The Netherlands: Kluwer.

Nunes, T., Schliemann, A. D., & Carraher, D. W. (1993). *Street mathematics and school mathematics.* Cambridge: Cambridge University Press.

Pimm, D. (1995). *Symbols and meanings in school mathematics.* London: Routledge.

Pollak, H. O. (1969). How can we teach applications of mathematics? *Educational Studies in Mathematics, 2,* 393–404.

Pollak, H. O. (1987). Cognitive science and mathematics education. : A mathematician's perspective. In A. H. Schoenfeld (Ed.), *Cognitive science and mathematics education* (pp. 253–264). Hillsdale, NJ: Lawrence Erlbaum Associates.

Polya, G. (1957). *How to solve it.* Princeton, NJ: Princeton University Press.

Puchalska, E., & Semadeni, Z. (1987). Children's reactions to verbal arithmetical problems with missing, surplus or contradictory data. *For the Learning of Mathematics, 7 (3),* 9–16.

Radatz, H. (1983). Untersuchungen zum Lösen eingekleideter Aufgaben. *Zeitschrift für Mathematik-Didaktik, 4,* 205–217.

Radatz, H. (1984). Schwierigkeiten der Anwendung arithmetischer Wissen am Beispiel des Sachrechnens. In: *Untersuchungen zum Mathematikunterricht* (Band 10, pp. 17–29). Bielefeld, Germany: Institut für Didaktik der Mathematik, Universität Bielefeld.

Reed, S. K. (1999). *Word problems: Research and curriculum reform.* Mahwah, NJ: Lawrence Erlbaum Associates.

Renkl, A. (1999, August). *The gap between school and everyday knowledge in mathematics.* Paper presented at the Eighth European Conference for Research on Learning and Instruction, Göteborg, Sweden.

Resnick, L. B. (1987). Learning in school and out. *Educational Researcher, 16 (9),* 14–21.

Resnick, L. B., Briars, D., & Lesgold, S. (1992). Certifying accomplishments in mathematics: The New Standards examining system. In I. Wirtzup & R. Streit (Eds.), *Developments in school mathematics around the world, Vol. 3,* pp. 186–207. Reston, VA: National Council of Teachers of Mathematics.

Restivo, S. (1993). The Promethean task of bringing mathematics to earth. In S. Restivo, J. P. Van Bendegem, & R. Fischer (Eds.), *Math worlds: Philosophical and social studies of mathematics and mathematics education* (pp. 3–17). Albany, NY: State University of New York Press.

Reusser, K. (1988). Problem solving beyond the logic of things: Contextual effects on understanding and solving word problems. *Instructional Science, 17,* 309–338.

Reusser, K. (1990). From text to situation to equation: Cognitive simulation of understanding and solving mathematical word problems. In H. Mandl, E. De Corte, N. Bennett, & H. F. Friedrich (Eds.), *Learning and instruction, Vol. 2* (pp. 477–498). Oxford: Pergamon.

Reusser, K., & Stebler, R. (1997a). Every word problem has a solution: The suspension of reality and sense-making in the culture of school mathematics. *Learning and Instruction, 7,* 309–328.

Reusser, K., & Stebler, R. (1997b, August). *Realistic mathematical modeling through the solving of performance tasks.* Paper presented at the Seventh European Conference on Learning and Instruction, Athens, Greece.

Riley, M. S., Greeno, J. G., & Heller, J. I. (1983). Development of children's problem-solving ability in arithmetic. In H. P. Ginsburg (Ed.), *The development of mathematical thinking* (pp. 153–196). New York: Academic Press.

Rogoff, B. (1984). Introduction: Thinking and learning in context. In B. Rogoff & J. Lave (Eds). (1984). *Everyday cognition* (pp. 1–8). Cambridge, MA: Harvard University Press.

Rogoff, B., & Lave, J. (Eds.). (1984). *Everyday cognition.* Cambridge, MA: Harvard University Press.

Romberg, T. A. (Ed.). (1995). *Reform in school mathematics and authentic assessment.* Albany, NY: State University of New York Press.

Romberg, T. A., Wilson, L., Khaketla, M., & Chavarria, S. (1992). Curriculum and test alignment. In T. A. Romberg (Ed.), *Mathematics assessment and evaluation: Imperatives for mathematics education* (pp. 21–38). Washington, DC: American Association for the Advancement of Science.

Ross, S. M. (1983). Increasing the meaningfulness of quantitative material by adapting context to student background. *Journal of Educational Psychology, 75,* 519–529.

Salomon, G. (Ed.). (1993a). *Distributed cognitions: Psychological and educational considerations.* New York: Cambridge University Press.

Salomon, G. (1993b). No distribution without individuals' cognition: A dynamic interactional view. In G. Salomon (Ed.), *Distributed cognitions: Psychological and educational considerations* (pp. 111–138). New York: Cambridge University Press.

Salomon, G., & Perkins, D. N. (1998). Individual and social aspects of learning. In P. D. Pearson & A. Iran-Nejad (Eds.), *Review of research in education, Vol. 23,* pp. 1–24. Washington, DC: American Educational Research Association.

Säljö, R. (1991). Learning and mediation: Fitting reality into a table. *Learning and Instruction, 1,* 261–273.

Säljö, R., & Wyndhamn, J. (1987). The formal setting as context for cognitive activities. An empirical study of arithmetic operations under conflicting premises for communication. *European Journal of Psychology of Education, 2,* 233–245.

Säljö, R., & Wyndhamn, J. (1988a). A week has seven days. Or does it? On bridging linguistic openness and mathematical precision. *For the Learning of Mathematics, 8 (3),* 16–20.

Säljö, R., & Wyndhamn, J. (1988b). Cognitive operations and educational framing of tasks: School as a context for arithmetical thought. *Scandinavian Journal of Educational Research, 32,* 61–71.

Säljö, R., & Wyndhamn, J. (1990). Problem-solving, academic performance, and situated reasoning: A study of joint cognitive activity in the formal setting. *British Journal of Educational Psychology, 60,* 245–254.

Säljö, R., & Wyndhamn, J. (1993). Solving everyday problems in the formal setting: An empirical study of the school as context for thought. In S. Chaiklin & J. Lave (Eds.), *Understanding practice: Perspectives on activity and context* (pp. 327–342). Cambridge: Cambridge University Press.

Saxe, G. (1988). Candy selling and math learning. *Educational Researcher, 17 (6),* 14–21.

Schank, R. C. (1982). *Reading and understanding.* Hillsdale, NJ: Lawrence Erlbaum Associates.

Schank, R. C., & Abelson, R. P. (1977). *Scripts, plans, goals and understanding.* Hillsdale, NJ: Lawrence Erlbaum Associates.

Schoenfeld, A. H. (1985). *Mathematical problem solving.* Orlando, FL: Academic Press.

Schoenfeld, A. H. (1987a). What's all the fuss about metacognition? In A. H. Schoenfeld (Ed.), *Cognitive science and mathematics education* (pp. 61–88). Hillsdale, NJ: Lawrence Erlbaum Associates.

Schoenfeld, A. H. (1987b). Confessions of an accidental theorist. *For the Learning of Mathematics, 7 (1),* 30–38.

Schoenfeld, A. H. (1988). When good teaching leads to bad results: The disasters of "well-taught" mathematics courses. *Educational Psychologist, 23,* 145–166.

Schoenfeld, A. H. (1991). On mathematics as sense-making: An informal attack on the unfortunate divorce of formal and informal mathematics. In J. F. Voss, D. N. Perkins, & J. W. Segal (Eds.), *Informal reasoning and education* (pp. 311–343). Hillsdale, NJ: Lawrence Erlbaum Associates.

Schoenfeld, A. H. (1992). Learning to think mathematically: Problem solving, metacognition, and sense making in mathematics. In D. A. Grouws (Ed.), *Handbook of research on mathematics teaching and learning* (pp. 334–370). Macmillan: New York.

School Mathematics Study Group (1970). Problem formulation. In *Secondary school mathematics: Unit 2. Teacher's commentary* (pp. 47–67). Stanford, CA: Author.

Scieszka, J., & Smith, L. (1995). *Math curse.* New York: Viking.

Scribner, S. (1984). Studying working intelligence. In B. Rogoff & J. Lave (Eds.), *Everyday cognition: Its development in social context* (pp. 9–40). Cambridge, MA: Harvard University Press.

Selter, C. (1994). How old is the captain? *Strategies, 5 (1),* 34–37.

Semadeni, Z. (1995). Developing children's understanding of verbal arithmetical problems. In M. Hejný & J. Novotná (Eds.), *Proceedings of the International Symposium on Elementary Math Teaching* (pp. 27–32). Prague, The Czech Republic: Faculty of Education, Charles University.

Silver, E. A. (Ed.). (1985). *Teaching and learning mathematical problem solving. Multiple research perspectives.* Hillsdale, NJ: Lawrence Erlbaum Associates.

Silver, E. A. (1986). Using conceptual and procedural knowledge: A focus on relationships. In J. Hiebert (Ed.), *Conceptual and procedural knowledge: The case of mathematics* (pp. 181–189). Hillsdale, NJ: Lawrence Erlbaum Associates.

Silver, E. A. (1994). On mathematical problem posing. *For the Learning of Mathematics, 14 (1),* 19–28.

Silver, E. A., & Kenney, P. A. (1995). Sources of assessment information for instructional guidance in mathematics. In T. Romberg (Ed.), *Reform in school mathematics and authentic assessment* (pp. 38–86). Albany, NY: State University of New York Press.

Silver, E. A., Shapiro, L. J., & Deutsch, A. (1993). Sense making and the solution of division problems involving remainders: An examination of middle school students' solution processes and their interpretations of solutions. *Journal for Research in Mathematics Education, 24,* 117–135.

Sowder, J. (1992). Estimation and number sense. In D. A. Grouws (Ed.), *Handbook of research on mathematics teaching and learning* (pp. 371–389). New York: Macmillan.

Sowder, L. (1988). Children's solutions of story problems. *Journal of Mathematical Behavior, 7,* 227–238.

Sowder, L. (1989). Searching for affect in the solution of story problems in mathematics. In D. B. McLeod & V. M. Adams (Eds.), *Affect and mathematical problem solving: A new perspective* (pp. 104–113). New York: Springer.

Srinivasiengar, C. N. (1967). *The history of ancient Indian mathematics.* Calcutta, India: The World Press.

Staub, F. C., & Reusser, K. (1995). The role of presentational structures in understanding and solving mathematical word problems. In C. A. Weaver, S. Mannes, & C. Fletcher (Eds.), *Discourse comprehension: Essays in honour of Walter Kintsch* (pp. 285–305). Hillsdale, NJ: Lawrence Erlbaum Associates.

Stern, E. (1992). Warum werden Kapitänsaufgaben "gelöst"? Das Verstehen von Textaufgaben aus psychologischer Sicht. *Der Mathematikunterricht, 28 (5),* 7–29.

Stewart, I. (1987). *The problems of mathematics.* Oxford: Oxford University Press.

Stewart, I., & Golubitsky, M. (1992). *Fearful symmetry: Is God a geometer?* London: Penguin.

Stigler, J. W., Fuson, K. C., Ham, M., & Kim, M. S. (1986). An analysis of addition and subtraction word problems in American and Soviet elementary mathematics textbooks. *Cognition and Instruction, 3,* 153–171.

Stinissen, J., Mermans, M., Tistaert, G., & Vander Steene, G. (1985). *Leuvense Schoolvorderingentest Vernieuwde Wiskunde 2–6* [Leuven Standard Achievement Test New Mathematics 2–6]. Brussels, Belgium: Centrum voor Studie- en Beroepsoriëntering.

Streefland, L. (1988). Reconstructive learning. In A. Borbas (Ed.), *Proceedings of the Twelfth International Conference for the Psychology of Mathematics Education, Vol. 1,* pp. 75–91. Veszprem, Hungary: OOK Printing House.

Swan, M. (1993). Improving the design and balance of mathematical assessment. In M. Niss (Ed.), *Investigations into assessment in mathematics education* (pp. 195–215). Dordrecht, The Netherlands: Kluwer.

Swetz, F. J. (1987). *Capitalism and arithmetic: The new math of the 15th century.* La Salle, IL: Open Court.

Thompson, A. (1992). Teachers' beliefs and conceptions: A synthesis of the research. In D. A. Grouws (Ed.), *Handbook of research on learning and teaching mathematics* (pp. 127–146). New York: Macmillan.

Thorndike, E. L. (1922). *The psychology of arithmetic.* New York: Macmillan.

Toom, A. (1999). Word problems: Applications or mental manipulatives. *For the Learning of Mathematics, 19 (1),* 36–38.

Treffers, A. (1987). *Three dimensions: A model of goal and theory description in mathematics education: The Wiskobas project.* Dordrecht, The Netherlands: Reidel.

Treffers, A., & De Moor, E. (1990). *Proeve van een nationaal programma voor het reken-wiskundeonderwijs op de basisschool: Deel 2. Basisvaardigheden en cijferen* [Towards a national mathematics curriculum for the elementary school: Part 2. Basic skills and written computation]. Tilburg, The Netherlands: Zwijssen.

Van den Heuvel-Panhuizen, M. (1996). *Assessment and realistic mathematics education.* Utrecht, The Netherlands: University of Utrecht.

Van Haneghan, J. P., Barron, L., Young, M. F., Williams, S. M., Vye, N. J., & Bransford, J. D. (1992). The Jasper series: An experiment with new ways to enhance mathematical thinking. In D. F. Halpern (Ed.), *Enhancing thinking skills in the sciences and mathematics* (pp. 15–38). Hillsdale, NJ: Lawrence Erlbaum Associates.

Van Lieshout, E. C. D. M., Verdwaald, A., & Van Herk, J. (1997, August). *Suppression of real-world knowledge and demand characteristics in word problem solving.* Paper presented at the Seventh European Conference on Learning and Instruction, Athens, Greece.

Van Oers, B. (1998). From context to contextualising. *Learning and Instruction, 8,* 473–488.

Vergnaud, G. (1982). A classification of cognitive tasks and operations of thought involved in addition and subtraction problems. In T. P. Carpenter, J. M. Moser, & T. A. Romberg (Eds.), *Addition and subtraction: A cognitive perspective* (pp. 39–59). Hillsdale, NJ: Lawrence Erlbaum Associates.

Vergnaud, G. (1996). The theory of conceptual fields. In L. P. Steffe, P. Nesher, P. Cobb, G. A. Goldin, & B. Greer (Eds.), *Theories of mathematical learning* (pp. 219–239). Mahwah, NJ: Lawrence Erlbaum Associates.

Verschaffel, L. (1979). *Kwalitatief-psychologische analyse en beïnvloeding van het probleemoplossend denken. Een onderzoek met elementaire rekenopgaven bij 6–8–jarige kinderen* [A qualitative-psychological analysis and improvement of problem solving. An investigation with 6–8–year old children solving elementary arithmetic problems]. Unpublished Master's dissertation: University of Leuven, Belgium.

Verschaffel, L. (1984). *Representatie- en oplossingsprocessen van eersteklassers bij aanvankelijke redactie-opgaven over optellen en aftrekken. Een theoretische en methodologische bijdrage op basis van een longitudinale, kwalitatief-psychologische studie* [First graders' representation and solution processes of elementary addition and subtraction word problems]. Unpublished doctoral dissertation, University of Leuven, Belgium.

Verschaffel, L. (1999). Realistic mathematical modelling and problem solving in the upper elementary school: Analysis and improvement. In J. H. M. Hamers, J. E. H. Van Luit, & B. Csapó (Eds.), *Teaching and learning thinking skills* (pp. 215–240). Lisse, The Netherlands: Swets & Zeitlinger.

Verschaffel, L., & De Corte, E., (1993). A decade of research on word-problem solving in Leuven: Theoretical, methodological, and practical outcomes. *Educational Psychology Review, 5,* 239–256.

Verschaffel, L., & De Corte, E. (1996). Number and arithmetic. In A. J. Bishop, K. Clements, C. Keitel, J. Kilpatrick, & C. Laborde (Eds.), *International handbook of mathematics education, Vol. 1,* pp. 99–137. Dordrecht, The Netherlands: Kluwer.

Verschaffel, L., & De Corte, E. (1997a). Word problems. A vehicle for authentic mathematical understanding and problem solving in the primary school? In T. Nunes & P. Bryant (Eds.), *Learning and teaching mathematics: An international perspective* (pp. 69–98). Hove, England: Psychology Press.

Verschaffel, L., & De Corte, E. (1997b). Teaching realistic mathematical modeling and problem solving in the elementary school. A teaching experiment with fifth graders. *Journal for Research in Mathematics Education, 28,* 577–601.

Verschaffel, L., De Corte, E., & Borghart, I. (1997). Pre-service teachers' conceptions and beliefs about the role of real-world knowledge in mathematical modelling of school word problems. *Learning and Instruction, 4,* 339–359.

Verschaffel, L., De Corte, E., & Lasure, S. (1994). Realistic considerations in mathematical modelling of school arithmetic word problems. *Learning and Instruction, 4,* 273–294.

Verschaffel, L., De Corte, E., & Lasure, S. (1999). Children's conceptions about the role of real-world knowledge in mathematical modeling of school word problems. In W. Schnotz, S. Vosniadou, & M. Carretero (Eds.), *New perspectives on conceptual change* (pp 175–189). Oxford: Elsevier.

Verschaffel, L., De Corte, E., Lasure, S., Van Vaerenbergh, G., Bogaerts, H., & Ratinckx, E. (1999). Design and evaluation of a learning environment for mathematical modeling and problem solving in upper elementary school children. *Mathematical Thinking and Learning, 1,* 195–229.

Verschaffel, L., De Corte, E., & Vierstraete, H. (1999). Upper elementary school pupils' difficulties in modeling and solving non-standard additive word problems involving ordinal numbers. *Journal for Research in Mathematics Education, 30,* 265–285.

Victorian Curriculum and Assessment Board (1990). *Mathematics study design.* Carlton, Victoria: Author.

Vye, N. J., Schwartz, D. L., Bransford, J. D., Barron, B., Zech, L., & The Cognition and Technology Group at Vanderbilt (1997). SMART environments that support monitoring, reflection, and revision. In D. J. Hacker, J. Dunlosky, & A. C. Graessar (Eds.), *Metacognition in educational theory and practice* (pp. 305–346). Mahwah, NJ: Lawrence Erlbaum Associates.

Vygotsky, L. S. (1986). *Thought and language.* Cambridge, MA: MIT Press.

Wells, D. (1992). *The Penguin book of curious and interesting puzzles.* London: Penguin.

Wells, D. (1997). *The Penguin book of curious and interesting mathematics.* London: Penguin.

Weyl, H. (1969). *Symmetry.* Princeton, NJ: Princeton University Press.

Whimbey, A., Lochhead, J., & Potter, P. (1990). *Thinking through math word problems.* Hillsdale, NJ: Lawrence Erlbaum Associates.

Wittgenstein, L. (1956). Remarks on the foundations of mathematics. (G. H. von Wright, R. Rhees, & G. E. M. Anscombe, Eds., and G. E. M. Anscombe, Trans.). Oxford: Blackwell.

Wyndhamn, J., & Säljö, R. (1997). Word problems and mathematical reasoning: A study of children's mastery of reference and meaning in textual realities. *Learning and Instruction, 7,* 361–382.

Yackel, E., & Cobb, P. (1996). Sociomathematical norms, argumentation, and autonomy in mathematics. *Journal for Research in Mathematics Education, 27,* 458–477.

Yoshida, H., Verschaffel, L., & De Corte, E. (1997). Realistic considerations in solving problematic word problems: Do Japanese and Belgian children have the same difficulties? *Learning and Instruction, 7,* 329–338.

Name Index

Subject Index

CONTEXTS OF LEARNING
Classrooms, Schools and Society

1. *Education for All.* Robert E. Slavin
   1996. ISBN 90 265 1472 7 (hardback)
   ISBN 90 265 1473 5 (paperback)

2. *The Road to Improvement: Reflections on School Effectiveness.* Peter
   Mortimore
   1998. ISBN 90 265 1525 1 (hardback)
   ISBN 90 265 1526 X (paperback)

3. *Organizational Learning in Schools.* Edited by Kenneth Leithwood and
   Karen Seashore Louis
   1999. ISBN 90 265 1539 1 (hardback)
   ISBN 90 265 1540 5 (paperback)

4. *Teaching and Learning Thinking Skills.* Edited by J.M.H. Hamars, J.E.H.
   van Luit and B. Csapó
   1999. ISBN 90 265 1545 6 (hardback)

5. *Managing Schools Towards High Performance: Linking School
   Management Theory to the School Effectiveness Knowledge Base.* Edited
   by Adrie J. Visscher
   1999. ISBN 90 265 1546 4 (hardback)

6. *School Effectiveness: Coming of Age in the Twenty-First Century.* Pam
   Sammons
   1999. ISBN 90 265 1549 9 (hardback)
   ISBN 90 265 1550 2 (paperback)

7. *Educational Change and Development in the Asia-Pacific Region: Challenges
   for the Future.* Edited by Tony Townsend and Yin Cheong Cheng
   2000. ISBN 90 265 1558 8 (hardback)
   ISBN 90 265 1627 4 (paperback)

8. *Making Sense of Word Problems.* Lieven Verschaffel, Brian Greer and Erik
   de Corte
   2000. ISBN 90 265 1628 2 (hardback)